AF718854

Geology of the Arisaig Area, Antigonish County, Nova Scotia

with articles by

Arthur J. Boucot
Department of Geology, Oregon State University, Corvallis, Oregon

John F. Dewey
Department of Geological Sciences, State University of New York, Albany, New York

David L. Dineley
Department of Geology, The University, Bristol, England

Raymond Fletcher
Department of Geology, Brown University, Providence, Rhode Island

William K. Fyson
Department of Geology, University of Ottawa, Ottawa, Ontario

John G. Griffin
Department of Geology, University of California, Davis, California

Charles F. Hickox
Department of Geology, Colby College, Waterville, Maine

W. Stuart McKerrow
Department of Geology and Mineralogy, Oxford University, Oxford, England

Alfred M. Ziegler
Department of Geophysical Sciences, University of Chicago, Chicago, Illinois

THE GEOLOGICAL SOCIETY OF AMERICA

3300 Penrose Place • Boulder, Colorado 80301

SPECIAL PAPER 139

Library of Congress Catalog Card Number 72-82614
I.S.B.N. 0-8137-2139-3

Published by
THE GEOLOGICAL SOCIETY OF AMERICA, INC.
3300 Penrose Place
Boulder, Colorado 80301

Printed in the United States of America
by Edwards Brothers, Inc., Ann Arbor, Michigan 48104.
Type composed by Auto-Graphics, Inc.,
Monterey Park, California 91754.

The printing of this volume
is made possible through the bequest of
Richard Alexander Fullerton Penrose, Jr.

Contents

FIGURE

TABLE

Acknowledgments

The authors are indebted to C. W. Harper, Jr., for the use of material from his doctor's thesis on the brachiopods of the Arisaig Group. In that study, he identified many of the small collections that have helped to make correlation possible between the shore and the inland areas.

W.B.N. Berry and M. J. Copeland have greatly assisted by identifying the graptolites and the ostracods, respectively, found at Arisaig. R. K. Bambach has generously contributed his knowledge of the Arisaig bivalves; he has also permitted examination of Twenhofel's field notebook.

The chapter on the Knoydart has greatly benefited from the work of R. H. Denison, D. J. MacNeil, A. Suparman, and M. Tauchid.

John F. Dewey gratefully acknowledges financial support from the University of Manchester, the Royal Society of London, and the Nova Scotia Research Foundation.

D. Kelley has been generous with information on the Cobequid Mountains.

We are indebted for funds received by W. S. McKerrow from the Nova Scotia Research Foundation, the Royal Society of London, and the Geological Society of London Tyrrell Fund, and by D. L. Dineley from the Geological Survey of Canada and the National Museum of Canada.

McKerrow spent the year 1964-1965 as a post-doctoral Research Fellow at the California Institute of Technology with support from the National Science Foundation (Grants GP-2290 and GP-3743).

Acknowledgment is made in the list of maps to those Massachusetts Institute of Technology students who contributed to the brook maps presented in this report, but thanks are due to many others who assisted with the mapping of the terrain between the brooks. David Ziegler deserves special thanks for locating small outcrops in the bush.

Finally, we all are deeply appreciative of the thorough review given the manuscript by Marland P. Billings.

Abstract

The richly fossiliferous Silurian and Early Devonian rocks of the Arisaig area and the nearby related regions of Cape George and Lochaber in Antigonish County, Nova Scotia, have been known for almost one hundred and fifty years. The 5,000 ft of sediment in the Arisaig Group is the only fossiliferous section within the Province containing representatives of every stage of the Silurian. This monotonous succession of siltstones and mudstones is unique world-wide in providing an almost continuous faunal record of very shallow water conditions rather than an alternation of varying depth environments. The *Eocoelia* and related benthic animal communities persist from the early Llandovery through the entire Silurian and well into the early Gedinnian portion of the Lower Devonian.

These rocks despite their relatively unmetamorphosed nature (never higher than the chlorite grade) have been highly folded and extensively faulted, beginning with disturbances associated with the Middle Devonian Acadian orogeny and extending into various intervals of the later Paleozoic and possibly Triassic. The complex structure is here outlined in more detail than is available in previous reports, thus giving both geologists and paleontologists a better framework with which to understand both faunal and geologic problems associated with the rocks. During the Gedinnian, the entire region became a site of nonmarine, Old Red Sandstone–type deposition, but there is no evidence for any stratigraphic break between the underlying marine and the overlying nonmarine beds. The Devonian vertebrates are similar to those of the Welsh Borderland of England.

The rhyolitic and basaltic volcanics underlying the marine Silurian rocks of the Arisaig area are described in some detail for the first time. These volcanic rocks were extruded into a nonmarine environment. The precise age of the volcanics is still uncertain, although a Late Ordovician or Early Silurian age is reasonable.

The Arisaig succession is discussed in relation to the paleogeography of the adjacent part of the Appalachians.

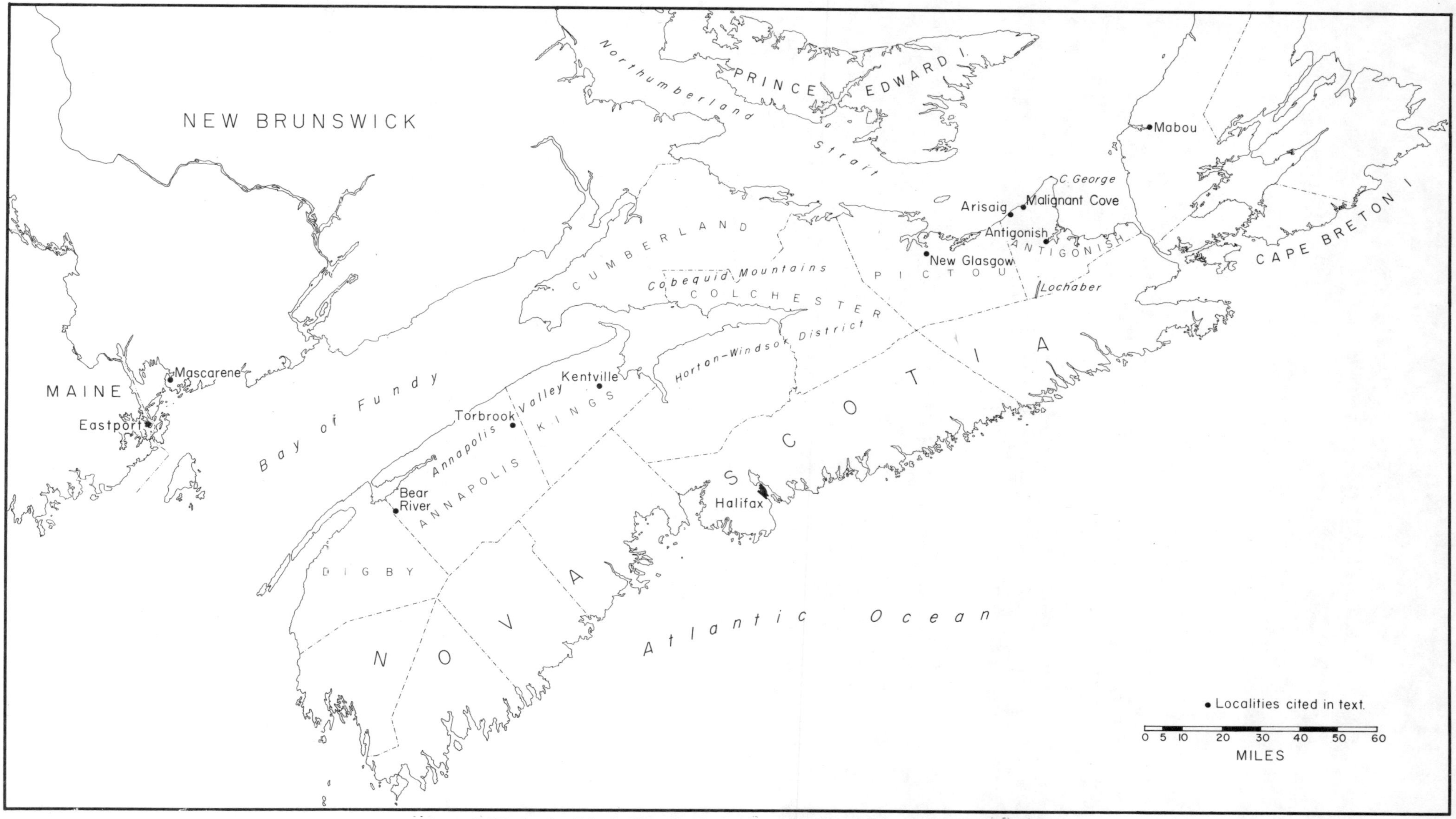

Figure 1. Silurian localities in Nova Scotia, southern New Brunswick, and southeast Maine.

BOUCOT AND OTHERS, FIGURE 1
Geological Society of America Special Paper 139

Introduction

PURPOSE AND SCOPE OF THE INVESTIGATION

The Arisaig area (Fig. 1; Pl. 1) has been selected for detailed study because it provides the most continuous and best exposed sections of marine Silurian and early Lower Devonian rocks in the Appalachian Mountain system. The earliest fossiliferous rocks present are lower Llandovery in age, and it is probable that the whole of the Silurian is represented in the 4,000 ft of beds that make up the Arisaig Group; but the underlying Bears Brook Volcanic Group has no known fossils, and the Silurian rocks are faulted against the Ordovician Browns Mountain Group. It is not clear, therefore, how large a stratigraphic break, if any, is present at the base of the Arisaig Group. The marine Silurian succession passes without a break up into marine Lower Devonian; the system junction comes near the top of the Stonehouse Formation. The basal marine Devonian beds are followed conformably by the nonmarine Knoydart Formation (about 1,000 ft thick) which contains Lower Devonian (upper Downtonian to middle Dittonian) fish remains.

Special studies have also been made of the Bears Brook Volcanic Group and of the structure of the area; repeated movements along the Hollow Fault have resulted in complex folding and faulting in the inland parts of the area.

The village of Arisaig (Fig. 1) lies on Northumberland Strait, 15 mi northeast of Antigonish, Nova Scotia. It is now easily accessible from New Glasgow or Antigonish by Nova Scotia Route 45. The area studied extends along the coast from McAras Brook (3 mi southwest of Arisaig) to McKinnons Brook (3 mi to the northeast of Arisaig); the area is bounded to the southeast by the Hollow Fault, a major transcurrent fault topographically marked by a well-defined linear depression, which is between 1 and 2 mi inland in this region. The coast provides almost continous exposures in cliffs and in the tidal zone. The coast sections have been the subject of several investigations, but no previous workers have mapped the formations inland in detail, although good exposures also occur along many deeply incised brooks.

Much of the complex structure in the vicinity of the Hollow Fault has been worked out by collecting from all known fossil horizons; collections (mainly brachiopods) have been obtained from 146 localities, 36 on the shore and 110 inland (*see* Appendix). One result of the inland mapping has been to show that some formations and parts of others are faulted out on the shore sections.

A brief account of the Arisaig Group rocks in the inland area near Lochaber, about 25 mi south-southeast of Arisaig, is presented. In addition, the geology of the Cape George region, about 15 mi northeast of Arisaig, is outlined in order to provide a background for an account of the Arisaig Group and related Silurian rocks found in two fault zones present near the tip of Cape George.

PREVIOUS GEOLOGICAL WORK

McLearn (1924) gives a full account of the history of geological work in the Arisaig area. Gesner (1836) was the first to mention the Arisaig rocks, and Dawson (1845, 1850, 1855, 1891) described the strata and some of the fossils which he thought were of Silurian or Devonian age. Hall (*in* Dawson, 1855) first considered the beds to be Devonian, and then (1860) of Silurian age. This change of mind was no doubt the result of Honeyman's (1859) comparison of the Arisaig fauna with the Silurian faunas of Britain. Subsequently, Honeyman (1864, 1867, 1870, 1886, 1890) subdivided the Arisaig rocks into four stratigraphic units; he thought his lowest subdivision (A) was "Lower Silurian," that is, Ordovician in modern terminology, and his highest subdivision (D) was upper Ludlow or lower Helderberg.

Fletcher (1887) gave the first detailed description of the Arisaig beds; he accepted Honeyman's correlations, but did not list fossil occurrences. Fletcher's maps (1893) do not agree in detail with the present interpretation of the area. Ami (1900) designated four units: Stonehouse, Moydart, McAdam, and Arisaig, but gave neither diagnostic fossils nor exact lithological boundaries. Ami also named and described the Knoydart formation (1900b) and considered its contact with the Stonehouse formation (1901).

Twenhofel (1909) divided the rocks of the Arisaig coast section into 40 lithological zones and listed the fossils present in each. His classification is the first that permits recognition in the field; it is summarized in Table 1.

It is probable that Twenhofel's formations do not correspond exactly with Ami's original units, but subsequent authors have all used Twenhofel's formation boundaries, although with some changes in nomenclature. Twenhofel and Schuchert (*in* Twenhofel, 1909) considered the ages of these divisions to range from lower Llandovery to Ludlow.

Williams (1914) published a map, the inland parts of which were not based on fossil evidence, but he did point out in his memoir that the beds containing the iron ore in Arisaig Brook are at the base of his McAdam Formation and are faulted out on the shore. Williams also renamed Twenhofel's Division I "the Beechhill Cove Formation" and Division II "the Ross Brook Formation"; he stated that he and Twenhofel examined the formation boundaries in the field, and used them as Twenhofel delimited them. Williams used the term Arisaig Series for the five formations: Stonehouse, Moydart, McAdam, Ross Brook, and Beechhill Cove. It is now renamed "Arisaig Group."

McLearn (1924) gave a good account of all the faunas known at that time, but his main collections came from the coastal section and his map is only a slightly modified version of that of Williams (1914). He was correct in placing the Beechhill Cove Formation in the lower Llandovery, in recognizing upper Llandovery forms in the Ross Brook Formation, and in recording the lower Ludlow graptolite *Monograptus wandalensis* in the McAdam Formation. Earlier (1918), McLearn had discussed the problems raised by the absence of many common Silurian faunas in the Arisaig Group.

By means of fossils and lithology, he was able to subdivide most of Williams' five formations; many of McLearn's subdivisions are here recognized as members (Table 1).

O. T. Jones (1926) collected graptolites at the base of the Ross Brook Formation which he considered to be early upper Llandovery in age. He also collected later upper Llandovery graptolites from Twenhofel's zone 8.

In 1958, Boucot and others at the M.I.T. summer field camp started a program of mapping the coast and brooks of the Arisaig area on a scale of 50 ft to the inch; these maps form the basis of the maps in the present work. At the same time, Maehl (1961), under Boucot's supervision, was working in the Paleozoic area of eastern Pictou County, between 10 and 20 mi southwest of Arisaig. Maehl recognized the same sequence of formations in the northeast part of his area, except that he defined some 175 ft of beds between the Ross Brook Formation and the McAdam Formation (which then appeared to be absent at Arisaig) as a new French River Formation. Further, in line with the work already proceeding at Arisaig, Maehl subdivided many of the formations of the Arisaig Group into

TABLE 1. Previous and Present

Period	Series	Geological Age	Zone	Group	Present work (Arisaig)	Maehl (1961) (Northeast Pictou Co.)
DEVONIAN		Lower and middle Dittonian and upper Downtonian		ARISAIG GROUP	KNOYDART FM.	KNOYDART FM.
		Lower Gedinnian			STONEHOUSE FM.	STONEHOUSE FM.
SILURIAN	Upper	Pridoli				
		Ludlow			MOYDART FM.	MOYDART FM.
					Red member	Upper Member
					Green member	Lower Member
					McADAM BROOK FM.	McADAM BROOK FM.
					Upper member	Upper Member
					Lower member	Upper Member
					Lower member	Upper Member
		Wenlock			DOCTORS BROOK FM.	Lower Member
					FRENCH RIVER FM.	FRENCH RIVER FM.
		Upper Llandovery	C_6		ROSS BROOK FM.	ROSS BROOK FM.
					Upper member	Upper Member
					Upper member	Upper Member
			C_1-C_5		Middle member	Middle Member
					Middle member	Middle Member
					Middle member	Middle Member
					Middle member	Middle Member
					Middle member	Middle Member
	Lower	Middle Llandovery			Lower member	"Lower Member"
		Lower Llandovery			BEECHHILL COVE FM.	BEECHHILL COVE FM.
?ORDOVICIAN					BEARS BROOK VOLCANIC GROUP	BEAR BROOK FM.
					Malignant Cove Fm.	
					McGillivray Brook Fm.	
					Dunn Point Fm.	
					BROWNS MOUNTAIN GROUP	BROWNS MOUNTAIN GROUP

Stratigraphic Divisions of the Arisaig Group

MAEHL (1961) (SOUTHEAST PICTOU CO.)	MCLEARN (1924) (ARISAIG)	WILLIAMS (1914) (ARISAIG)	TWENHOFEL (1909) (ARISAIG)
KNOYDART FM.	KNOYDART FM.	KNOYDART FM.	KNOYDART FM.
STONEHOUSE FM.	STONEHOUSE FM.	STONEHOUSE FM.	Division IVb or STONEHOUSE FM.
	Zone d	————	————
	Zone c	Zone 12	Zones 39-40
	Zone b	Zone 12	Zones 36½-38
	Zone a	Zone 11	Zones 33-36½
KERROWGARE FM.	MOYDART FM.	MOYDART FM.	Division IVa or MOYDART FM.
	Zone b	Zone 10	Zone 32
	Zone a	Zone 9	Zones 28-31
KERROWGARE FM.	McADAM FM.	McADAM FM.	Division III or McADAM FM.
	Zone c	Zone 8	Zones 26-27
	Zone c	Zone 8	Zones 21-25
	Zone c	Zone 6	Zone 21
	Zone b	Zone 6	Zones 13-20
KERROWGARE FM.	Zone a	Zone 5	————
KERROWGARE FM.	ROSS BROOK FM.	ROSS BROOK FM.	Division II or ARISAIG FM.
	Zone e	Zone 4	Zones 11-12
	Zone d	Zone 4	Zones 8-10
	Zone d	Zone 3	Zone 7
	Zone c	Zone 3	Zone 6
	Zone b	Zone 3	Zone 5
	Zone b	Zone 2	Zones 3-4
	————	————	————
	Zone a	Zone 2	Zone 2
GLENCOE BROOK FM.	BEECHHILL FM.	BEECHHILL COVE FM. Zone 1	Division I Zone 1
SUNNYBRAE FM.	————	BEARS BROOK Malignant Cove FM.	————
CHARCOAL FM.	————	————	————
BROWNS MOUNTAIN GROUP	————	BROWNS MOUNTAIN GROUP	————

members (Table 1). The table also shows Maehl's formations in southeast Pictou County where the Arisaig succession is so different from the type area that new formations have been designated. Some mapping was also carried out by M.I.T. parties in the Cape George area, 14 mi northeast of Arisaig (Boucot and others, 1959), and in the Lochaber area, 24 mi south of Arisaig.

Maehl has also carried out chemical analyses of many of the sediments of the Arisaig Group.

Benson (1970b) has mapped the Arisaig area on a scale of 1:50,000.

WORK BY PRESENT AUTHORS

Boucot made fossil collections during the summer of 1953. Boucot, Hickox, and Sage supervised field mapping by M.I.T. students during the summers of 1958–1960; further field mapping was done by McKerrow in 1962 and 1964, which included checking of all the M.I.T. maps; Ziegler, Dewey, and McKerrow are responsible for the maps of the Bears Brook Volcanic Group, and Dewey has contributed the section on these rocks. The section on structure is by Dewey and Fyson; the maps and sections showing the general geology are the work of Dewey, Fyson, and McKerrow. Dineley has contributed the section on the Knoydart Formation, and Hickox is responsible for the lithological descriptions of the Arisaig Group. McKerrow and Boucot are responsible for the Introduction, and most of the sections on General Geology, Geological History, Regional Relations, and the Acadian Orogeny. McKerrow and Hickox wrote the section on Physical Features. Responsibility for the Lochaber area and the Cape George area is outlined under the appropriate sections.

GEOLOGICAL SETTING

Arisaig lies in the northern Appalachian region where prolonged periods of subsidence during the Paleozoic were interrupted by orogenic events from time to time, notably in the Upper Ordovician and Middle Devonian; the rocks of the Bears Brook Volcanic Group and the Arisaig Group were emplaced between these two events.

Silurian and Lower Devonian rocks are known in many other areas of Nova Scotia (Fig. 1), to the south of Arisaig near Lochaber (Benson, 1964), to the southwest in Pictou County (Maehl, 1961), to the southwest in the Cobequid Mountains (Kelley, 1966; Dawson, 1878, supplement, p. 75), and in the Annapolis Valley (Smitheringale, 1960; Crosby,

1962). To the northeast, the nearest outcrops are at Cape George (Boucot and others, 1959; Benson, 1970a). Wait (1959) records Silurian sandstones containing rhynchonellids and mollusks west of the Mabou Highlands in Cape Breton Island, and Weeks (1954) considers unfossiliferous sandstones and conglomerates of the Middle River Group of southeast Cape Breton Island to be of Silurian age.

Berry and Boucot (1970) recognize Late Silurian facies belts distributed as in Figures 25 through 27. The Arisaig Group forms part of a shale and sandstone belt bounded to the northwest by the volcanics of southern New Brunswick and of Eastport and Penobscot Bay on the Maine coast; volcanics in Cape Breton may also be of the same age. To the southeast, the earlier Silurian sedimentary rocks are coarser than those at Arisaig, but the Late Silurian and Early Devonian beds become finer grained and pass into a shale belt, which is seen in the Annapolis Valley and South Pictou County (Smitheringale, 1960; Crosby, 1962).

The regular pattern of facies does not continue northeast of Arisaig if Weeks is correct in assigning the Middle River Group to the Silurian.

Except for the basal Beechhill Cove conglomerate, the Arisaig Group consists of fine-grained sedimentary rocks–silt-sized particles embedded in a matrix of clay minerals. This matrix has now been regionally altered to sericite and chlorite. The rocks are therefore low-grade metamorphics, even though they are treated as sedimentary rocks throughout this report.

The benthic invertebrate communities of the Arisaig Group all indicate very shallow-water conditions (Ziegler and Boucot, *in* Berry and Boucot, 1970). During the late Llandovery–early Wenlock, *Eocoelia* Community faunas were present. During the early Llandovery, later Wenlock, Ludlow, Pridoli, and early Gedinne, faunas equivalent in depth to the *Eocoelia* Community were present. The early Llandovery includes *Cryptothyrella* Community; the later Wenlock and younger beds, *Salopina* Community.

The brachiopod evidence and primary structures in some of the rocks (for example, channels and current bedding) strongly point to very shallow water deposition for the greater part of the succession. It seems probable, therefore, that the source of the sediments lay some tens of miles distant to the north and northwest, and for the lower part of the group, possibly also to the southeast. Sedimentation apparently kept pace with the subsidence in the basin of deposition so that shallow-water conditions persisted. The only times when conditions changed radically were during the deposition of the red member of the Moydart Formation and the Knoydart; both are nonmarine, but neither shows any significant increase in grain size.

Physical Features

TOPOGRAPHY AND SUPERFICIAL DEPOSITS

The shoreline northeast from Arisaig Point (Pl. 1) is controlled by the outcrop of the rocks of the Bears Brook Volcanic Group. Between the shore and the Hollow Fault, however, the physical features depend more on the distribution of glacial drift than on the structure of the Paleozoic rocks.

The Hollow Fault, as its name suggests, lies along a valley, the base of which is never more than 400 ft above sea level. It is bordered to the southeast by a steep scarp rising to about 800 ft within 1,500 ft of the valley. This scarp marks the margin of a plateau underlain by Ordovician rocks of the Browns Mountain Group. A few hills, formed by monzonite intrusions, rise above the undulating surface of the plateau.

Williams (1914, p. 41) and Goldthwait (1924, p. 28) describe the plateau as part of a widespread surface extending over the whole of the Maritime Provinces, the remains of this "Atlantic Upland" now being "preserved only on the more resistant rocks." Goldthwait follows Daly (1901) in considering the plateau to have been planed off in Cretaceous times; the plateau is certainly post-Triassic, but it could be much younger.

In describing the Arisaig region, Williams (1914, p. 44–49) divided the lowlands into "The Shore Front" and the "Carboniferous Lowland." "The Shore Front" covers the area of Silurian and Devonian rocks around Arisaig, where hills reach a height of 750 ft west of McAras Brook (Pl. 1) and gradually descend northeastward to 650 ft near the source of McAdam Brook, 500 ft east of Arisaig Brook, 250 ft near the source of McGillivray Brook, and 150 ft northeast of the lower reaches of Doctors Brook. This progressive decrease in height toward the northeast suggests a tilted erosion platform rather than erosion of the less resistant beds–especially as the most resistant beds, the basal volcanics, in the Arisaig area are those at lowest topographic level. It is thus possible that the "Shore Front" is a part of the Brown's Mountain Plateau faulted down to the northwest due to Tertiary movements along the Hollow Fault, and tilted so that the northeastern parts around Doctors Brook dropped more than did the McAras Brook area. Northeast of Malignant Cove, where the Hollow

TABLE 2. Previous and Present Topographic

Distance (ft) along coast between adjacent localities	Present usage	Copeland, 1964	National Topographic Series, 1953, 1:50,000
	Malignant Cove	——	same
6,000			
	McKinnons Brook	——	same
2,100			
	Doctors Brook	——	same
2,700			
	Dunn Point	——	same
1,200	Beechhill	——	——
	Cove Brook		
	Beechhill Cove	——	Beechhill Cove
4,400			
	McGillivray Brook	——	same
1,450	Frenchman's Barn	——	——
1,450	Mill Brook	——	Mill Brook
3,900			
1,200	Arisaig Point	——	same
	Ross Brook	——	——
2,350 (the mouth of Arisaig Brook is 300 ft SW. compared with before 1897)			
	Arisaig Brook	——	same
	Macdonald's Road	——	——
1,950	(as on signpost)		
400	Smith Brook	——	——
	French Brook	——	——
3,600	(new name)		
	McAdam Brook	same	——
2,350	Moydart Point	same	——(uses name
300			for point 3,300 ft SW.)
	McDonald Brook	same	——
1,500			
	Stonehouse Brook	same	——
700			
	MacEachern Brook	same	——
1,200			
	McAras Brook	same	McArras Brook
	MacPherson Brook	same	——
10,400	(northeast tributary of McAras Brook)		
	Knoydart Brook	——	same

N.B. "Same" means same name as column to left.
N.B. "——" means not mentioned.

Nomenclature of the Arisaig Area, Nova Scotia

McLearn, 1924	Williams, 1914	Geological Survey Dept. Canada 1893, 1:63,360
same	same	same
——	McNeil Brook	McNeil's Brook
Doctor Brook	same	Doctor's Brook
same	same	same
——	——	——
Beechhill Cove	Beechhill Cove	Beechhill Cove
Rory McDonald	——	Rory McDonald's
——	same	same
——	——(uses name as alternate for Knoydart Brook)	——
same	same	same
same	same	same
same	same	same
——	——	Trunk Road
same	Smiths Brook	Smith Brook
——	——	——
same	same	——
Moydart Point	same	——(uses name for point 500 ft SW.)
same	——	Joseph McDonald Brook
McPherson Brook	——	same (but head-waters shown draining into MacEachern Brook)
Stonehouse Brook (but 300 ft to NE. on his map)	——	Stonehouse Brook (but this brook does not now have head-waters southeast of highway)
McAras Brook	same	same
——	——	——
——	same	same

Fault runs along the shoreline, the surface of the downthrown northwestern block lies below sea level.

The latest movement in the Arisaig area was general uplift, probably associated with those sporadic, epeirogenetic movements that have affected northeastern North America intermittently since early Tertiary time. The general uplift is reflected by valleys deeply incised through glacial drift and into the underlying bedrock. Typical depths of incision (below the top of the drift) are: Doctors Brook (lower part), 100 ft; Doctors Brook, West Branch (in Hollow Fault), 250 ft; Arisaig Brook (at iron mine), 250 ft; Arisaig Brook (upper part, in Hollow Fault), 300 ft; and McAdam Brook (below Route 4A), 100 ft.

The two largest streams in the area, Doctors Brook and Knoydart Brook, have gentle gradients throughout most of their length; all others have waterfalls occurring wherever resistant rocks crop out.

Both of these large streams cross the subsequent valley of the Hollow Fault and are responsible for most of its drainage. The size of the valley, however, suggests that it was eroded prior to the last glacial epoch, and was postglacially exhumed by headward erosion of Doctors Brook and Knoydart Brook. This erosion beheaded the smaller consequent streams that flow northwestward into Northumberland Strait. Arisaig Brook, for example, is the most recent stream to have been beheaded by headward growth of Doctors Brook.

The drift, composed of till and glacio-fluvial sands, covers 85 percent of the area. Exposures along deeply incised streams indicate local thicknesses of at least 75 ft, and the absence of bedrock outcrops suggests that all but three of the hills in the map area are either thickly mantled by drift or are completely composed of drift. The three bedrock hills are the hill east of Arisaig Brook, the high land on both sides of lower McAdam Brook, and Browns Mountain, the Ordovician highland which forms the upthrown, southeastern block of the Hollow Fault.

The lack of raised beaches along the shore of Northumberland Strait indicates that the latest uplift predated the last glaciation, and that postglacial rebound, in response to unloading, must have been offset by postglacial eustatic rise in sea level. Drowned forests at Guysborough and Amherst indicate that sea level has continued to rise into the present time around Nova Scotia. This rise partly explains such submergent features as the deep embayments at Merigomish and Antigonish Harbors and the active wave-cutting of cliffs along the shore at Arisaig and Cape George. Only the larger streams in the area are graded to sea level; the mouths of the smaller ones are marked by waterfalls, indicating that cliff retreat is more rapid than stream downcutting.

TOPOGRAPHIC NOMENCLATURE

Considerable nomenclatural differences appear in the geological literature on the Arisaig area. Table 2 is a summary of the past and present nomenclature (*see also* Pls. 1 and 2). The present usage follows that of the local populace or that given in the 1953 National Topographic Series 1:50,000 map except for the following: Beechhill Cove–spelled with two h's, as in literature published before the 1953 map, and as used in designation of Beechhill Cove Formation by Williams (1914). Beechhill Cove Brook–name given to brook flowing into cove of this name. Macdonald's Road–name on signpost on Route 45 for road up the valley of Arisaig Brook. French Brook–name for brook 400 ft west of Smith Brook. Moydart Point–for the point designated by Williams (1914) and McLearn (1924) where beds of the Moydart Formation are well exposed. The 1953 map uses the term for a point formed by Mississippian rocks just north of McAras Brook.

TIDES AND TIDE MARKS

The normal tidal range at Arisaig is only 2 ft, and the water is lowest when steady offshore winds prevail, regardless of the state of the tide. It is thus meaningless to indicate low and high water marks on the detailed maps (Pls. 3 and 4).

Moreover, from year to year while this mapping was being carried out (1958 to 1964), the positions of sand, boulders, and rock outcrops changed considerably in many places. The shore maps, therefore, show all outcrops recorded, although they have not all been exposed simultaneously. The lowest edge of grass is also marked, but with a thinner line.

General Geology

DESCRIPTION OF FORMATIONS

Browns Mountain Group

The chloritized graywackes, slates, argillites, and hematite beds of this group occur on the southeast side of the Hollow Fault and were not examined closely during the present study.

Williams (1914, p. 55) records the inarticulate brachiopod *Obolus (Lingulobus) spissa* from iron-ore beds and associated micaceous sandstones on the East Branch of Doctors Brook and north of the Little Hollow (which runs parallel to the Hollow three-quarters of a mile southeast of the West Branch of Doctors Brook). The same brachiopod occurs with the Arenig (Lower Ordovician) graptolite *Didymograptus nitidus* in the iron-ore beds of Belle Isle, Newfoundland (Williams, 1914, p. 55, footnote). The probability of an Ordovician age is strengthened by the fact that the Browns Mountain Group is intruded by monzonite stocks which do not intrude the Silurian, and further, because pebbles from these stocks occur in the Malignant Cove Formation; the Browns Mountain Group is therefore certainly pre-Silurian, and probably older than the Late Ordovician or lower Llandovery Bears Brook Volcanic Group.

Browns Mountain conglomerates, with pebbles up to 1 cm, grits, and sandstones occur high on the east bank of Doctors Brook above stations L17 and L21 (Pls. 5A, 5B); and basic tuffs and serpentine are exposed in the Hollow Fault zone along the East Branch of Doctors Brook (stations U88 to U94, Pl. 5C). Other outcrops noted include grits on the West Branch of Doctors Brook 8,000 ft upstream from the junction of the East and West Branches, and epidotized basic rocks on the East Branch 3,500 ft upstream from the same junction (Pl. 1).

Bears Brook Volcanic Group

Maehl (1961, p. 29–31) gives reasons for including the rocks below the Beechhill Cove Formation under the term Bears Brook Formation; they are now designated the Bears Brook Volcanic Group. The rocks are superbly exposed on the coast between Arisaig Point in the southwest and Malignant Cove (Fig. 1) in the northeast. The Group consists of silicic and intermediate volcanic rocks of subaerial origin (the Dunn Point and McGillivray Brook Formations) and the fluviatile conglomerates of the Malignant Cove Formation (Fig. 2; Table 3). Although a slight unconformity exists at Arisaig beneath the Beechhill Cove Formation, it is impossible to directly assess its regional nature and significance. The regional strike and dip of the Bears Brook Group and the Beechhill Cove Formation are the same and, although no regional unconformity is apparent, the nature of the interval between the emplacement of the Bears Brook Group and the marine Llandoverian transgression is difficult to evaluate. Although the Bears Brook Group may be of lower Llandoverian or Upper Ordovician age, it is clear, from the large quantities of quartzite and chloritic graywacke contained in tuffs in the McGillivray Brook Formation and the pebbles in the Malignant Cove Formation, that it postdates the metamorphism of the Browns Mountain Group. Small quartz monzonite fragments also occur in the tuffs, which suggests that the quartz monzonite suite that intruded into the Browns Mountain Group also predates the Bears Brook Group.

South of the Hollow Fault zone, in the vicinity of Malignant Cove, an assemblage of coarse nonmarine red breccias and conglomerates (Malignant Cove Formation) has yielded the following pebbles: chloritic graywacke; sericitic quartzite; vein quartz; quartz monzonite; pink, white, purple, and buff pelites and feldspar porphyrys; green chert; jasper; spilite; spherulitic and flow-banded rhyolite; banded tuffs; green and purple welded tuffs, and purple flinty rhyolite. This clearly indicates that the Malignant Cove Formation postdates (a) the metamorphism of the Browns Mountains Group, (b) the intrusion of quartz monzonite, and (c) the volcanic formations of the Bears Brook Group. The precise relation of the Malignant Cove Formation to the Beechhill Cove Formation remains uncertain, but the uniformity of facies above the basal Beechhill Cove conglomerate suggests

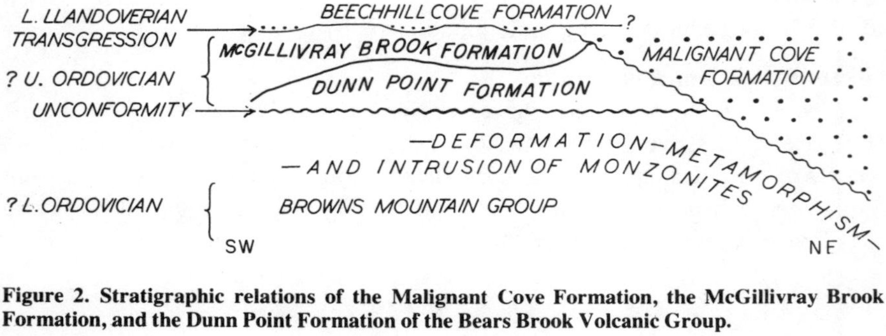

Figure 2. Stratigraphic relations of the Malignant Cove Formation, the McGillivray Brook Formation, and the Dunn Point Formation of the Bears Brook Volcanic Group.

TABLE 3. Paleozoic Rocks at Arisaig

	Period	Rock Units	Lithology	Thickness (ft)
MISSISSIPPIAN		Ardness Formation	Sandstone, shale, and limestone	2,045
		McAras Brook Formation unconformity	Conglomerate and grits, with basalt flows	1,145
LOWER DEVONIAN	(lower and middle Dittonian, and upper Downtonian)	Knoydart Formation	Red and green silty mudstones and marls, sandstones, and some interclast conglomerates	1,000 plus
SILURIAN UPPER	(Gedinnian and Pridoli)	Stonehouse Formation	Siltstones, shales, and mudstones; limestone lenses	1,350
		Moydart Formation	Green siltstone, limestone lenses, red marls, and concretionary limestones at top	390
	(Ludlow)	McAdam Brook Formation	Black shales and green siltstones, and some limestone concretions	350
	(Wenlock)	Doctors Brook Formation	Siltstones, shales, and silty limestones	275
	(Wenlock and upper Llandovery)	French River Formation	Siltstones with some shales, one hematite bed	280
SILURIAN LOWER	(middle and upper Llandovery)	Ross Brook Formation	Black shales with muddy siltstones and some arenaceous limestones, both increasing toward the top	1,050 to 1,300
	(lower Llandovery)	Beechhill Cove Formation	Siltstones, micaceous shales, sandstones, basal conglomerate in places	225 to 300
ORDOVICIAN ?UPPER		Malignant Cove Formation	Coarse nonmarine red breccias and conglomerates	
		McGillivray Brook Formation	Ignimbrites, lahars, rhyolites, and lateritic mudflows	
		Dunn Point Formation	Andesites, laterites, ignimbrites, rhyolites	
ORDOVICIAN ?LOWER		Browns Mountain Group	Argillite, slate, and graywacke, with some hematite beds	

that the fluviatile Malignant Cove Formation (Williams, 1914, p. 59) predates the marine beds of the Arisaig Group. The basal conglomerate of the Beechhill Cove Formation exposed in Doctors Brook contains a variety of pebbles similar to those of the Malignant Cove beds, and they could be of approximately the same age. But, in spite of the absence of an angular unconformity at the base of the Arisaig Group, it is thought that some considerable time interval may have elapsed after the emplacement of the volcanics and pyroclastic rocks and before the marine Llandoverian transgression, during which the Malignant Cove Formation was emplaced as a fluviatile wedge. The relations suggested above are summarized diagrammatically in Figure 2.

It is unlikely that the Llandoverian sea transgressed a surface of very strong relief, because of the general uniformity of thickness and facies in the Beechhill Cove Formation in the Arisaig area; although sporadically distributed thin basal pebble beds suggest slight local variations in relief. The total evidence, although largely indirect, thus suggests an Upper Ordovician age for the Bears Brook Volcanic Group.

Although the volcanics are generally excellently exposed, the parallelism of the coastline with their regional strike inhibits the simple elucidation of a three-dimensional synopsis of facies variation. The only considearble break in exposure is in Beechhill Cove where a block of Silurian rocks is displaced northwestward between two strike-slip faults. The volcanic sequence is characterized by an exceedingly varied assemblage of lavas and pyroclastics which are predominantly andesitic in the lower part and rhyolitic in the upper part of the Group. However, the top of the andesite sequence may not be regarded as representing a time plane because local interleafing with silicic lavas and tuffs occurs. Detailed correlation across breaks in exposure are rendered difficult by the complex and rapid facies variations. The base of the volcanic sequence is never seen and the top is exposed at only five localities, but it is noteworthy that the exposed thickness of andesites is consistently between 200 and 250 ft, suggesting that the base of the sequence (? resting on Browns Mountain Group) lies just out to sea, parallel to the coast.

It is not possible to erect an unequivocal time-rock stratigraphy for the whole of the Bears Brook Volcanic Group on account of incomplete exposure in a sequence characterized by rapid facies changes (Pl. 3). Lithologic sequences may be easily defined in any one section, however. These sequences are summarized in Figure 3, and both proven and suggested correlations are incorporated.

Although the sequence may be generally divided into andesites below and silicic lavas and pyroclastics above, it is clear (Fig. 3, sec. 5) that

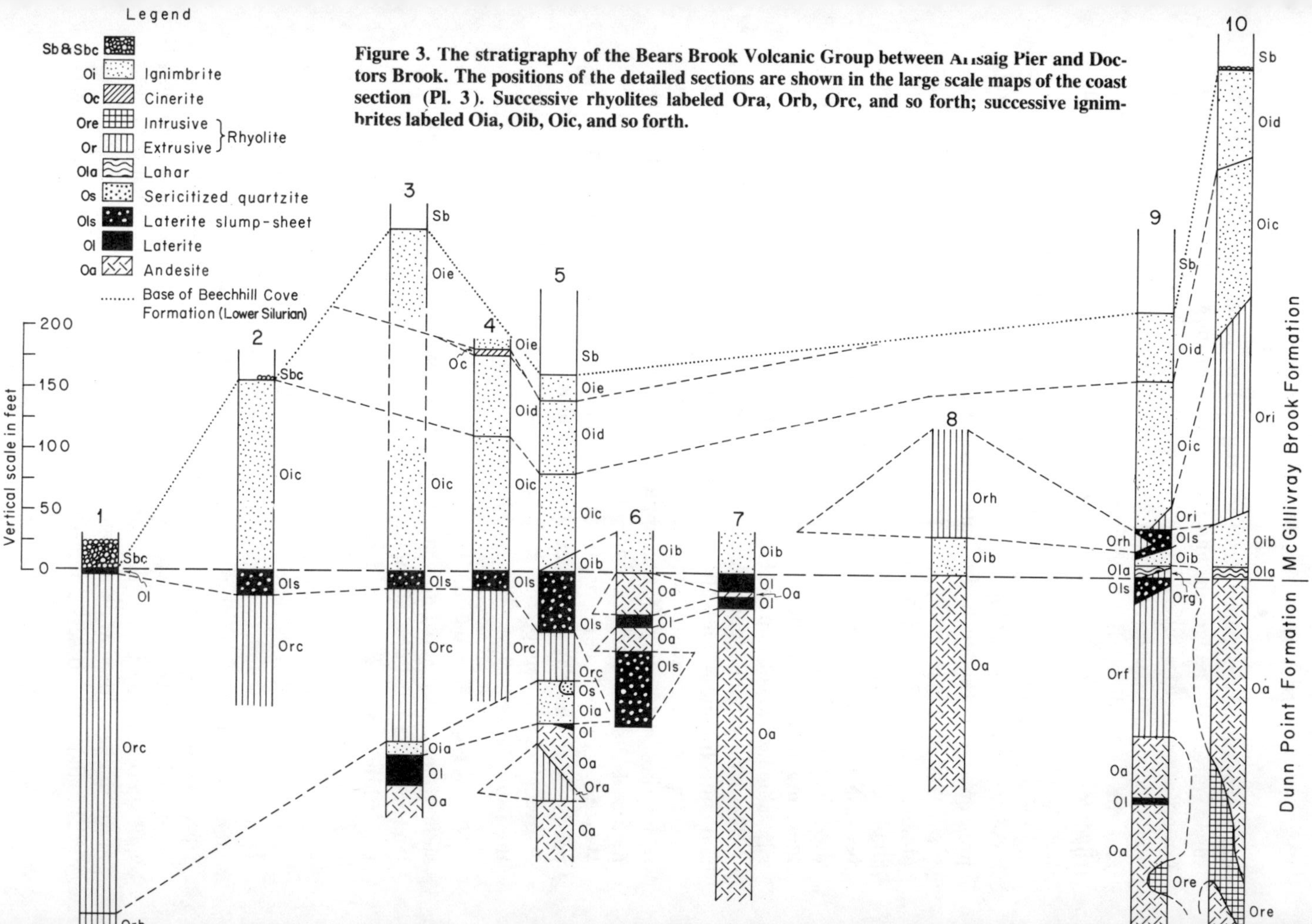

Figure 3. The stratigraphy of the Bears Brook Volcanic Group between Arisaig Pier and Doctors Brook. The positions of the detailed sections are shown in the large scale maps of the coast section (Pl. 3). Successive rhyolites labeled Ora, Orb, Orc, and so forth; successive ignimbrites labeled Oia, Oib, Oic, and so forth.

the extrusion of andesites partly overlapped in time the eruption of the silicic volcanics. The igneous and related rocks of the Bears Brook Volcanic Group are therefore divided (Fig. 3, sec. 5) into the Dunn Point Formation below, consisting of andesites, rhyolites, ignimbrites, and laterites, and the McGillivray Brook Formation above, consisting of silicic lahars, rhyolites, silicic ignimbrites, and laterites. The Malignant Cove Formation, containing pebbles of the other two formations is regarded as forming the top part of the Bears Brook Volcanic Group (Fig. 2).

West of section 5 (Fig. 3), three extrusive rhyolites (ra, rb, rc) are present as lens-shaped bodies, although in any one section no more than two rhyolites occur. Section 1 (Fig. 3), above the andesites, is composed almost wholly of extrusive rhyolites, but it is impossible to subdivide the thick rhyolite (rc) comprising the top of the Dunn Point Formation because there are gaps in exposure, and it may consist of more than one flow. Rhyolite "a" is present only in section 5 (Fig. 3). Unlike the rhyolites, the five ignimbrites (ia, ib, ic, id, ie) in the sequence, show considerable lateral constancy (Fig. 3, secs. 2 to 10).

The stratigraphic sequence is complicated by the intercalation of lahars and autochthonous and allochthonous laterites (Fig. 3, secs. 1 to 7, 9). The top of the andesite sequence is usually characterized by a thick bed of laterite, and locally (for example, sec. 6 on Fig. 3) considerable thicknesses of slumped laterite wedge out rapidly against andesites and rhyolites. West of section 5 (Fig. 3), a laterite marker horizon, partly bedded and partly slumped, may be traced almost continuously to Arisaig Point (Fig. 3, sec. 1); the top of this bed has been taken as the division between the Dunn Point and McGillivray Brook Formations.

Dunn Point Formation. The Dunn Point Formation (Fig. 3) is considered under two separate headings: first, the andesites and autochthonous laterites; secondly, the slumped laterites, ignimbrites, and rhyolites, here termed the Andesite Facies and Silicic Facies, respectively.

1. ANDESITE FACIES. The andesites attain a maximum thickness of 250 ft on the southwestern side of Malignant Cove where neither base nor top is seen. A surprising feature of the sequence is that in only one locality (between secs. 5 and 6, Fig. 3), is it possible to trace a flow into a feeder dike, and detailed mapping has shown that individual flows are rarely traceable over any great distance. Individual flows wedge out rapidly and clearly. The coast section cuts a series of lava tongues. Columnar jointing and sheet jointing is well developed in most flows and individual columns in the thickest flows attain a length of 6 ft.

The andesites possess a uniform and monotonous petrography consisting of plagioclase (An_{52}) phenocrysts and enstatite phenocrysts set in a pilotaxitic groundmass of plagioclase (An_{46}) laths and quartz. The plagioclase shows partial alteration to an aggregate of albite, epidote, and quartz, and the enstatite is commonly completely replaced by chlorite, serpentine, and iron ore. One thick flow contains very large plagioclase phenocrysts and forms a distinctive lithology traceable for a considerable distance on the shore, southwestward from Doctors Brook.

Most flows are highly vesicular, particularly at their tops, where large unoriented vesicles are filled with quartz, chlorite, and prehnite. Nearer the base, vesicles occur in distinct bands, and show a high degree of preferred orientation, indicating a northerly provenance for the flows.

The most distinctive and noteworthy feature of the andesites is that they commonly carry an undisturbed red lateritic soil profile. The vesicular tops of the flows pass up through a reddened zone of rubbly andesite into a laterite horizon. The simplest profile (Fig. 4) consists of a zone of a purple rubbly andesite-laterite mixture passing up gradually, or abruptly, into a blocky indurated red laterite. In thin section, the blocky laterite is seen to consist of a homogeneous aggregate of granular hematite, clay minerals, quartz, chalcedony, and a little calcite. Locally small nodules (perdigon) of carbonate and chalcedony occur and small white reduction patches occur sporadically near the top. The silica is clearly of two types; clear subangular quartz derived as a primary detrital mineral from the andesites and patches of irregular turbid quartz showing a system of irregular cracks. This latter variety is regarded as of secondary origin deposited in gel-form and produced, like the clay, as a product of hydrolysis. Some tiny enstatite grains and small epidotized plagioclase laths are present. Spherulitic texture is sparsely and poorly developed toward the base of the blocky red laterite.

The top of the profile commonly shows small irregular pockets infilled with bedded laterite, and this bedded fraction is regarded as having been

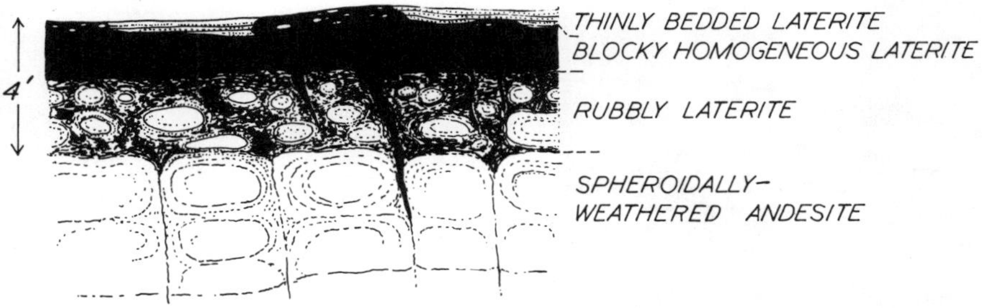

Figure 4. Laterite soil profiles developed on andesites of the Dunn Point Formation.

deposited by wind in wind-cut hollows. Some andesites are deeply weathered and reddened to depth well below the base of the distinct laterite profile. Columns show spheroidal weathering and deep reddening with liesegang banding and the intersections of columns with sheet joints result in their reduction to deeply weathered spheroidal boulders. A few thin flows are reduced to a mass of spheroidal blocks in a laterite matrix, and laterite dikes may penetrate the flows along joints, to a depth of 5 ft. Deep weathering along joints establishes a very useful criterion for distinguishing between cooling joints and later tectonic joints, as the former are characterized by a zone of reddening, whereas the latter cut weathered and fresh andesite alike and are lined by quartz and fresh green epidote.

In places, the lower parts of andesite flows have blebs and streaks of laterite derived from the underlying laterite. Trains of laterite inclusions may be traced directly from the base of a flow (Fig. 5A) where a penetrating tongue of laterite shows a gradual breakdown, first into blebs and then into streaks (Figs. 5B and 5C). Pull-apart and prolapsed structures are common in the inclusions, and in the few places where the sharp andesite-laterite contacts are destroyed, the laterite becomes broken down and incorporated in the andesite (Fig. 5D). Although trains of laterite inclusions may be traced directly into a basal flow cast, it is a common occurrence for a train to be sharply truncated by a slide surface near the base of, but within, the andesite. The intimate association of vesicles with the laterite trains indicates that the laterites were wet and plastic at the time of extrusion.

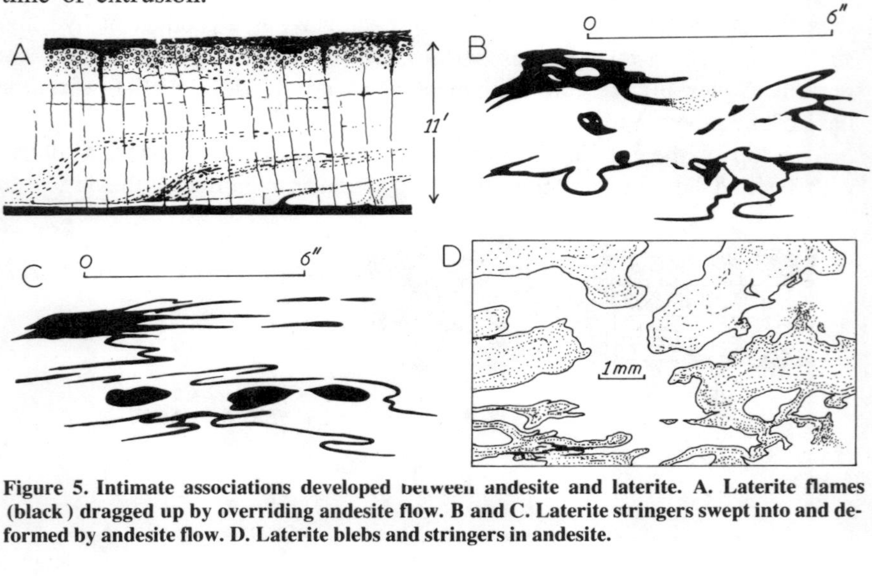

Figure 5. Intimate associations developed between andesite and laterite. A. Laterite flames (black) dragged up by overriding andesite flow. B and C. Laterite stringers swept into and deformed by andesite flow. D. Laterite blebs and stringers in andesite.

2. silicic facies. During and after the last phases of andesite extrusion, a period of deep lateritic weathering was accompanied by the cutting of deep channels in the andesites and the extrusion of rhyolite. In most areas, this episode is represented by a laterite horizon between the andesites and the McGillivray Brook Formation.

In Doctors Brook (Fig. 3, sec. 9, and Pl. 3), a pink and purple spherulitic rhyolite (rf), commonly showing flow banding, overlies the andesites and is overlain by a laterite. The top of the rhyolite is reddened and deeply weathered and the overlying laterite contains large angular blocks of weathered and reddened rhyolite. The top surface of the rhyolite shows grooving and fine striations, and it is suggested that the laterite was emplaced by sliding. The rhyolite is not present below the ignimbrite (ib) in section 8 of Figure 3, nor east of the Doctors Brook Fault (Fig. 3, sec. 10). A spherulitic rhyolite (re) occupying the fault zone near the mouth of Doctors Brook may be traced directly into the extrusive rhyolite (rf), and thence northwestward into a dike (re), gradually becoming more concordant with the andesite flows. Clearly the Doctors Brook Fault determined the distribution of the rhyolite and overlying laterite as indicated in Figure 6. The fault was also active during or after the emplacement of the McGillivray Brook Formation, because the latter shows marked thickness variation across the fault (Fig. 3, secs. 9 and 10; Pl. 3).

Between sections 3 and 7 (Fig. 3), a complex sequence of events has been established. During the period of deep lateritization, a northwest-southeast channel, at least 100 ft deep was cut in the andesites toward the top of the Dunn Point Formation. Then a flow-banded rhyolite tongue (ra) was emplaced along its southwestern flank (Fig. 7; Fig. 3, sec. 5, and

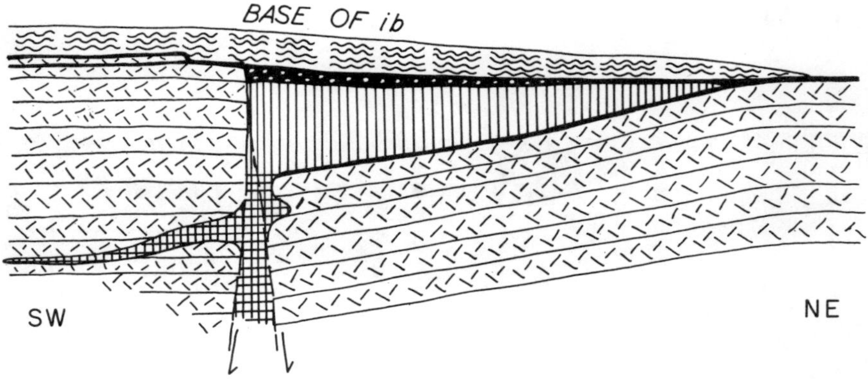

Figure 6. Structural and stratigraphic relations of the lowest extrusive rhyolite in Doctors Brook (ornament as in Fig. 3).

Pl. 3). The emplacement of the rhyolite was followed by a phase of laterite-slumping during which the channel was more than half-filled by a laterite containing spheroidal andesite blocks, subangular rhyolite fragments, and pellets of quartz-rich laterite. The complete infilling of the lower channel (Fig. 7) was effected by the extrusion of a thin andesite flow, the emplacement of a thin welded tuff (ia) and a thick extrusive rhyolite (rc), and the further accumulation of laterite. The extrusion of a thick andesite flow was followed by deep lateritization and the cutting of an upper channel (Fig. 7) into which slumped a mass of bouldery laterite containing andesite, rhyolite, and occasional quartzite pebbles. The slumped laterite in the upper channel thins rapidly to the northeast, although it continues as a horizon about 10 ft thick as far as section 3 (Fig. 3); the detailed stratigraphy of this horizon may be observed immediately to the east of Frenchman's Barn in section 4 of Figure 3. Here the top of the rhyolite (rc) is patchily reddened and is sharply truncated by a quartz-rich laterite with angular blocks of mottled purple-green rhyolite. Grooving on the surface of the rhyolite (rc) and small overturned folds in the laterite suggest slumping from the north-northwest. A further sharp break separates the lower slumped laterite from a laterite above which contains decomposed spheroidal andesite blocks. The blocky structure of the laterite and a continuous gradation from laterite to unweathered andesite in the cores of the blocks suggest that it was emplaced by sliding, with little internal distortion, onto the lower laterite. The upper portion of the laterite horizon consists largely of bedded and cleaved fine-grained quartz-rich laterite, suggesting emplacement by wind action. Farther to the southwest, the laterite horizon is bedded throughout and may be traced above the rhyolite (rc) to Arisaig (Fig. 3, sec. 1), where it is less than 1 ft thick.

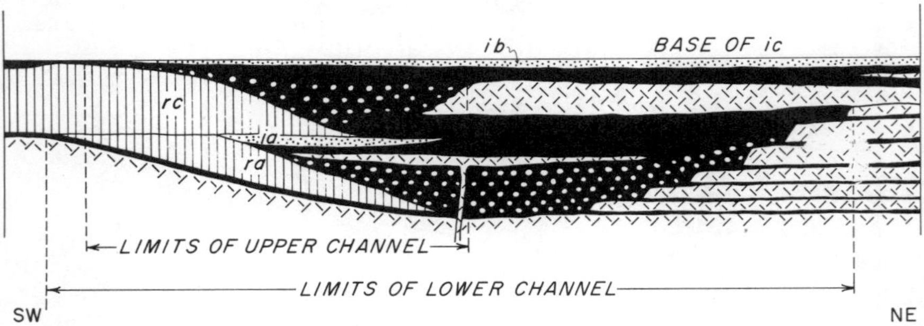

Figure 7. Diagrammatic reconstruction of the stratigraphy of the upper part of the Dunn Point Formation between sections 3 and 7 (ornament as in Fig. 3).

An interesting feature is that the lowest ignimbrite (ia) contains, near its top, a huge inclusion of sericitic quartzite (Fig. 3, sec. 5s). This lithology does not occur elsewhere in the volcanics of the Bears Brook Group as an integral stratigraphic unit but is exceedingly similar to the sericitic quartzite pebbles in the overlying Malignant Cove Formation. It is probably a block derived from the Browns Mountain Group that may have been carried by the incandescent ash cloud which was emplaced to form the ignimbrite (ia). The quartzite fragments in the laterite infilling the upper channel were also probably derived from the same source.

The full paleogeographic significance of the allochthonous laterites is fully discussed below.

McGillivray Brook Formation. The formation, consisting of an assemblage of ignimbrites, lahars, rhyolites, and lateritic mudflows, is fully exposed in Doctors Brook (Pl. 3; Fig. 3, secs. 9 and 10) and near the mouth of McGillivray Brook (Fig. 3, sec. 5). To the northeast, the lowest member is a blue-green argillite consisting of occasional epidotized plagioclase grains set in a matrix of clay minerals and bright-green chlorite. The rock possesses an irregular lenticular structure, and small tight overturned folds are common. The rock was clearly emplaced by slumping (overturned folds suggest a northerly provenance) and is interpreted as a lahar (la).

The lahar is succeeded in section 10 (Fig. 3) by a compact eutaxitic welded tuff (ib) which overlaps the lahar between sections 8 and 9 (Fig. 3) and onto the andesites. Near the mouth of Mill Brook (Pl. 3; Fig. 3, sec. 2) "ib" is equally compact and flinty but is characterized by the development of nodular texture. This ignimbrite (ib) extends as a discrete horizon to the northeast of Frenchman's Barn (Pl. 3; Fig. 3, sec. 4), where it consists of a thin sericitic sillar with small laterite inclusions.

Although "ib" is succeeded by "ic" near McGillivray Brook (Pl. 3; Fig. 3, sec. 5), a considerable time interval between emplacements of the two units is indicated by the presence of a thick extrusive rhyolite (rh) north of Beechhill Cove (Pl. 3, sec. 8), which thins to zero below a slumped laterite at Doctors Brook (Pl. 3, sec. 9).

Complete sections of the thickest and best developed of the ignimbrites (ic and id) are seen east of Doctors Brook (Pl. 3, sec. 10) and southwest of McGillivray Brook (Pl. 3, secs. 4 and 5). In section 10 (Pl. 3), they show little variation from a compact eutaxitic welded tuff. Considerable lithologic and textural variations are shown, however, around Frenchman's Barn and McGillivray Brook (Pl. 3; Fig. 3, secs. 4, 5, and 6); these merit detailed description.

The base of "ic" in section 4 consists of a grayish-red (5R 4/2)[1] microbreccia showing a general alignment of inclusions in a poorly foliated matrix. In thin section (Fig. 8), the matrix is seen to consist of a mass of unresolvable clay minerals, minute delicate undistorted devitrified glass shards, and granular hematite. The inclusions consist of fragments of welded and molded tuff, small brown pumice fragments, irregular subangular quartz, small andesite fragments, pellets of laterite and large plates of devitrified parataxitic and tubular pumice. The large pumice plates are irregular, show frayed ends and are broken and veined by the mesostasis.

The intimate association of delicate divitrified glass shards and granular derived hematite, together with the broken and veined pumice plates, suggests that the nuée ardente responsible for the emplacement of "ic" plowed into and incorporated a mass of unconsolidated laterite. The inclusion of large amounts of laterite presumably accelerated the de-gassing of the lower part of the flow and prevented welding and the development of

[1]Colors are given using the Rock-Color Chart published by the Geological Society of America, New York, 1951.

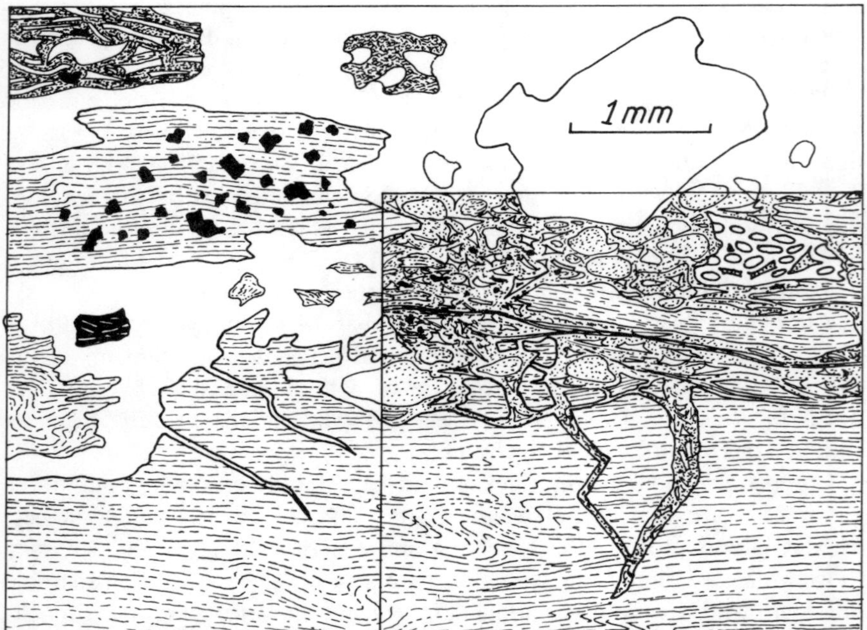

Figure 8. Drawing of a thin section of the basal laterite/ash carpet facies of the ignimbrite "ic" in section 5. Rounded and subrounded laterite pellets ornamented by stipple.

a eutaxitic foliation. The basal laterite-ash carpet facies of "ic" gradually changes to the southwest until, near Frenchman's Barn (Pl. 3; Fig. 3, sec. 3), the large pumice plates are no longer veined by the matrix and the hematite content is somewhat lower. Here the basal facies of the bed (ic) consists of a grayish-red (5R 4/2), poorly and irregularly foliated microbreccia. In thin section (Fig. 9), the matrix is seen to consist of fine delicate triangular and platy-shaped shards, showing pectinate devitrification, with numerous intact and undistorted single and multiple bubbles set in an irresolvable mesostasis of clay minerals and granular hematite. Laterite pellets are larger and less numerous than they are in section 4 (Fig. 3), and many contain a great deal more granular hematite. Fragments of sericitic welded tuff identical to "ic" are common, and pumice occurs as large elongate parataxitic plates and as undistorted fragments. It is clear, therefore, that the parataxitic texture of the larger plates was not developed by *in situ* vesicle collapse but by gas-streaming at the source vent. A great deal of the undistorted devitrified pumice shows recrystallization to a coarse strain-free quartz-mosaic, although preserving the vesicular texture of the pumice.

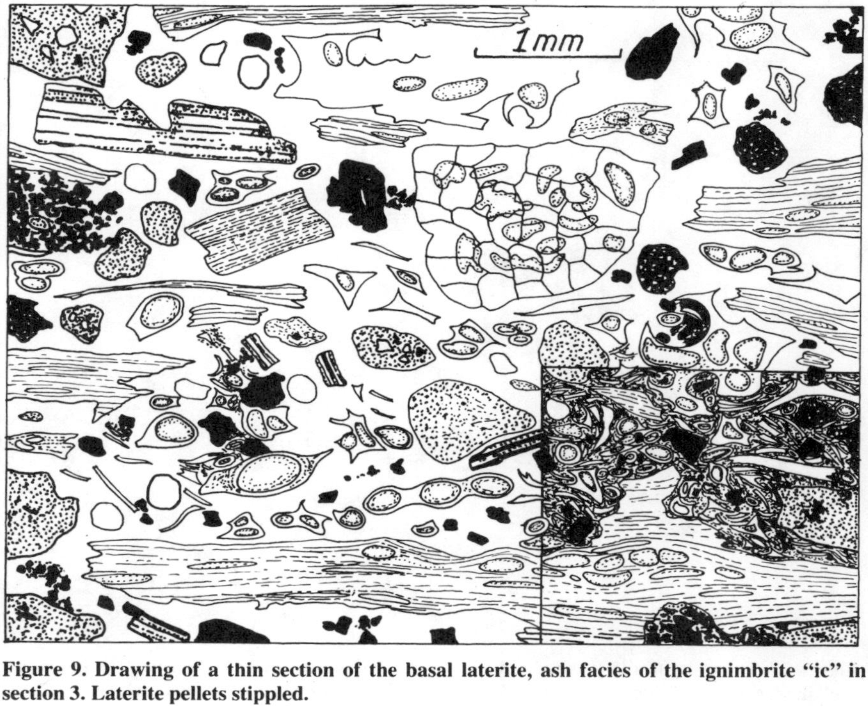

Figure 9. Drawing of a thin section of the basal laterite, ash facies of the ignimbrite "ic" in section 3. Laterite pellets stippled.

The basal laterite-ash carpet facies of "ic" grades rapidly upward into a grayish-brown (5YR 3/2), compact tuff in which a variety of undistorted inclusions are set in a granular quartz, feldspar, clay, and hematite mesostasis with devitrified glass shards and bubbles (Fig. 10). The shards and slightly flattened bubbles show a crude alignment, and slight molding of the foliation around the larger inclusions occurs. No welding of shards occurs in spite of the apparent eutaxitic foliation produced by shard alignment. The usual andesite and laterite inclusions are present with parataxitic pumice fragments, but the laterite generally possesses spherulitic texture.

This facies changes rapidly upward into a mottled (grayish-brown 5YR 3/2 to moderate brown 5YR 3/4) compact, strongly foliated flinty tuff with a distinctive pinkish-gray (5YR 8/1) weathering patina. The mottling, although partly irregular, occurs as streaks parallel to foliation; the cause is seen in thin section (Fig. 11) to result from the presence or absence of laterite pellets.

Glass shards and bubbles show distortion, flattening, and molding around inclusions. However, in the absence of any signs of welding, it is

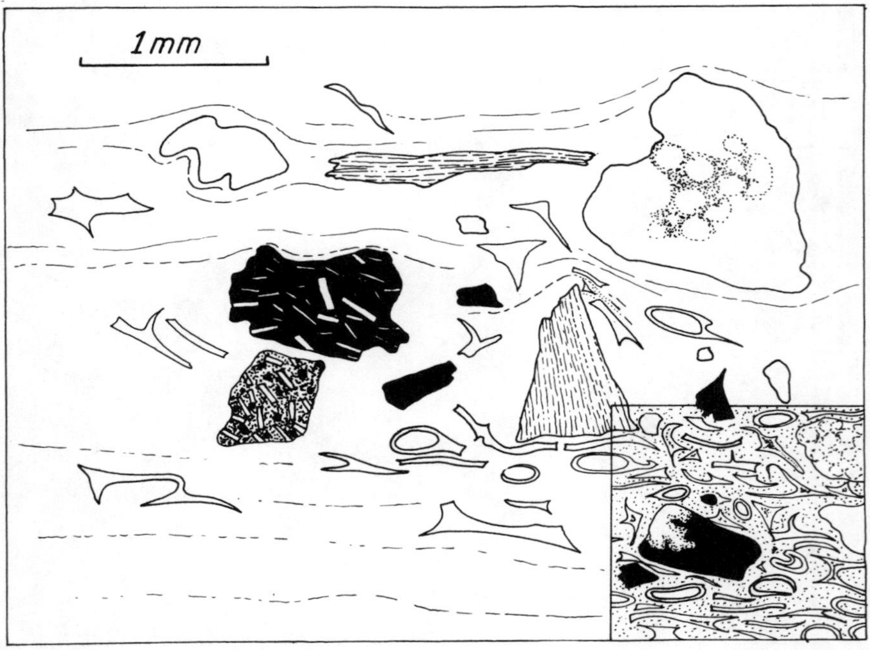

Figure 10. Drawing of a thin section of the lower partially molded tuff facies of the ignimbrite "ic" in section 5. Note large laterite inclusion (top right) showing spherulitic texture.

difficult to decide exactly how much the alignment of shards or flattening of shards has contributed to the development of the strong foliation. Euhedral quartz and subrounded fragments of ignimbrite are the most common inclusions, and both show bottle-necked embayments of the mesostasis. The origin of bottle-necked embayments has been discussed in some detail by Rast (1963), who concludes that the embayments develop *in situ* after emplacement of the hot ash flow because glass shards (as in the present example) have clearly been forced into the embayments. If the embayments existed in cognate or accidental inclusions during the ejection of the gas-ash cloud, it is difficult to understand why they were unfilled, and remained unfilled, during transport.

The foliated molded-tuff facies passes up rapidly into a grayish-red (10R 4/2), exceedingly well-foliated eutaxitic welded tuff (Fig. 12). The rock contains very little disseminated hematite, but large andesite fragments possess a corona of clay minerals presumably developed as a result of pneumatolytic action. Bottle-neck embayments in subrounded quartz are very common, and large shredded pumice fragments are ubiquitous. The plate-like elongate shape of the pumice is clearly not the result of

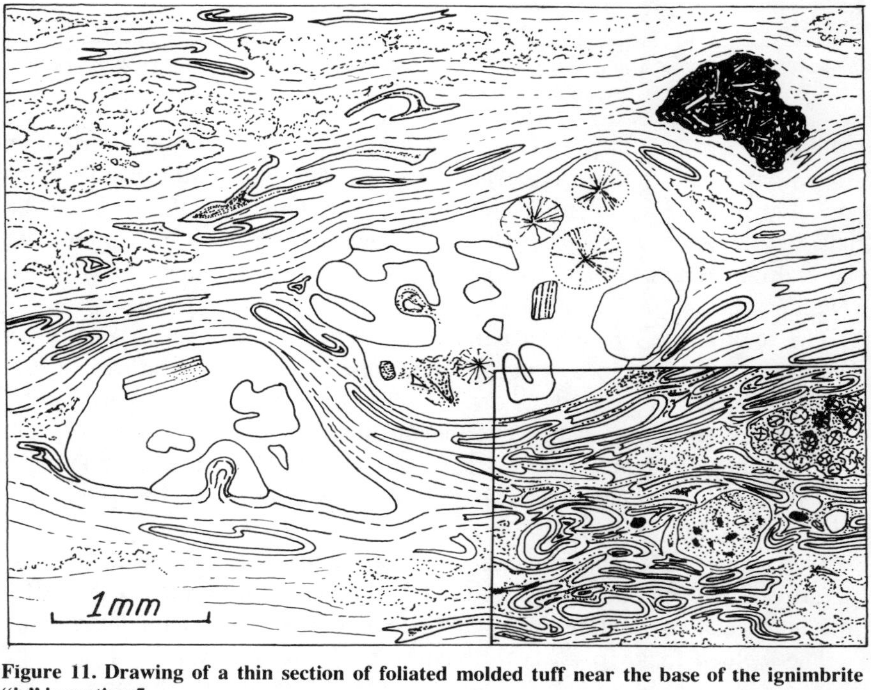

Figure 11. Drawing of a thin section of foliated molded tuff near the base of the ignimbrite "ic" in section 5.

Figure 12. Drawing of a thin section of the welded facies of the ignimbrite "ic" in section 5.

collapse *in situ,* as large quartz-filled vesicles are entirely undistorted. A strong parataxitic texture is developed in discrete bands, and in this compact flinty lithology, spherulites are developed. This gives the rocks in the field the appearance of a flow-banded spherulitic rhyolite and emphasizes the caution necessary in distinguishing between rhyolite lavas and welded rhyolite tuffs.

The rock is cut by numerous narrow chloritic tuffisite veins, a common feature in many welded and molded tuffs (Beavon and others, 1961; Dewey, 1963). These veins, however, clearly partly predate the eutaxitic foliation because where they cut the uncollapsed pumice plates, they are almost planar; whereas in the eutaxitic matrix, they are ptygmatically folded. This is in direct contrast to previously described tuffisite veins which cut the eutaxitic foliation.

The transition from the basal laterite ash member into eutaxitic and parataxitic welded tuff occurs through a thickness of between 5 and 10 ft and is directly correlatable with diminishing derived clay-hematite content.

The welded facies persists throughout the lower two-thirds of "ic" and then passes gradually back into an unwelded sillar facies. The upper sillar

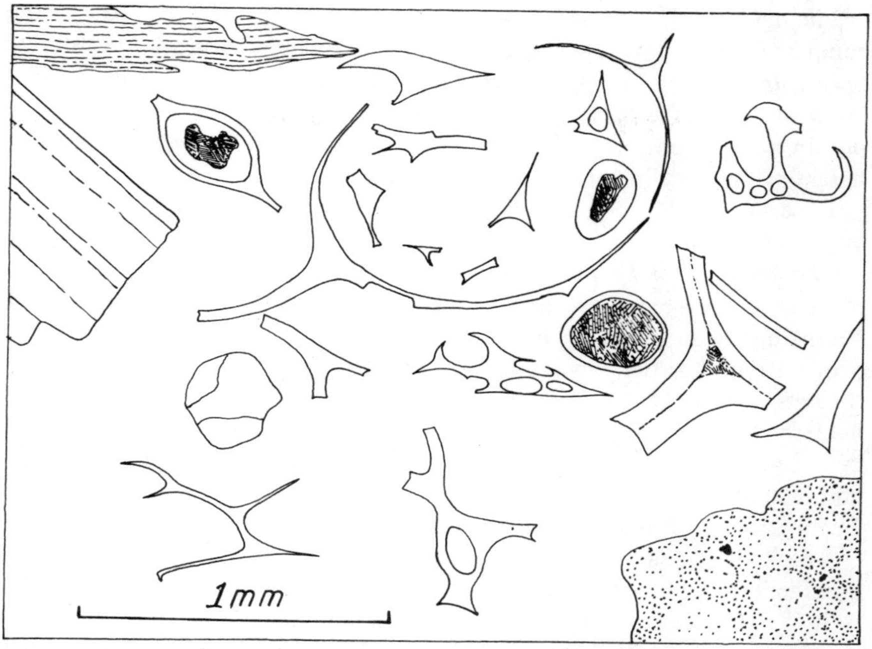

Figure 13. Drawing of a thin section of the upper sillar facies of the ignimbrite "ic" in section 5.

facies is a mottled (pale red, 10R 6/2 to dark yellowish-brown 10YR 4/2) somewhat flinty microbreccia. Plate-like parataxitic pumice is common and subrounded fragments of spherulitic laterite are occasionally seen (Fig. 13). The matrix consists of finely divided clay minerals, quartz, and feldspar with exceedingly delicate axiolitic shards and numerous chlorite-filled unbroken bubbles. The shards show no orientation, and there are no signs of molding around inclusions or welding. Sericitic quartzite and chloritic cleaved quartzite fragments are common and were probably derived from the Browns Mountain Group. Occasional quartz-rich laterite fragments show spherulitic texture and occasionally are recrystallized to a coarse quartz mosaic.

In Section 5 (Fig. 3), the ignimbrite "ic" is succeeded by a slump sheet consisting of an intermix of ash and laterite showing complete transitions from sharp to diffuse boundaries. The laterite fraction is clearly brecciated and tongued by the ash and later liesegang banding characterizes the whole slump sheet. The origin of the sheet may be ascribed to the plowing of an ash avalanche into an unconsolidated laterite accumulation. Derivation from the north is suggested by small prolapsed folds.

The ignimbrite "id" consists throughout of a grayish-red (10R 4/2) compact flinty eutaxitic welded tuff with parataxitic texture locally developed against collapsed and flattened pumice lapilli. The rock contains a large amount of laterite pellets which, although flattened and streaked out in some cases, are generally little deformed. This emphasizes that flattening and welding *may* occur under only slight vertical load.

In section 4 (Fig. 3), the ignimbrite "id" is succeeded by a thin-banded cinerite (c) which is in turn succeeded by "ie."

Malignant Cove Formation. Along Malignant Brook, 4½ mi northeast of Arisaig, conglomerates and coarse red sandstones crop out for about ¾ mi upstream from the junction of the road to Georgeville with Nova Scotia Route 45. Bedding is obscure, but there is probably a much greater thickness than the 20 ft estimated by Williams (1914, p. 58–59). No field relations are seen with other Ordovician or Silurian rocks, but the beds contain graywacke pebbles derived from the Browns Mountain Group and ignimbrite and rhyolite pebbles from the earlier formations of the Bears Brook Volcanic Group. It is therefore younger than these.

The beds appear to be of fluviatile origin, and Williams (1914, p. 59) considers the whole sequence to be terrestrially deposited.

It is not clear how this formation is related to the basal Beechhill Cove conglomerates; both contain pebbles of the Bears Brook volcanics, but they are overlain by different beds; the Malignant Cove conglomerates have red sandstones above them, the Beechhill Cove conglomerates are followed by marine siltstones.

The most probable explanation is that the marine Beechhill Cove Formation is later than the terrestrial Malignant Cove beds (Fig. 2). This explanation is supported by the evidence from Pictou County (Maehl, 1961, p. 30) where conglomerates similar to those at Malignant Cove underlie the Beechhill Cove Formation.

Arisaig Group

This group includes the beds from the lower Llandovery Beechhill Cove Formation to the Pridoli and lower Gedinnian Stonehouse Formation; its total thickness at Arisaig is about 4,000 ft (Table 3).

Beechhill Cove Formation. Williams (1914, p. 63) named this formation which is equivalent to Twenhofel's (1909) Division I (Table 1). At the type section in Beechhill Cove, 2 mi northeast of Arisaig, the beds lie on the volcanics of the Bears Brook Volcanic Group (Pl. 3). In places, a basal conglomerate is present, resting on the eroded surface of the volcanics, but there is no angular unconformity. A sharp contact is present at the top of the formation; the green shales and siltstones are distinct from the black shales of the overlying Ross Brook Formation.

DISTRIBUTION AND THICKNESS. The Beechhill Cove Formation is only present in the Arisaig area in the coastal region northeast of Arisaig Pier (Pl. 1); no exposures occur more than 1,000 ft from the coast. The dips are generally steep toward the southeast.

On the northeast side of Beechhill Cove, 300 ft of this formation is exposed; even though the base is still covered at low tide, this is the thickest sequence visible. Complete sections are exposed in Doctors Brook (Pl. 3A) and on the shore 650 ft southwest of Mill Brook (Pl. 3C); in both, the thickness is only 225 ft. Good exposures are present in McGillivray Brook, but (unlike the three sections mentioned above) the section is faulted and the exact thickness is uncertain. Thicknesses are exact only to within 20 percent, because of doubts as to the amount of structural deformation, but the variations in thickness may be due to deposition on an uneven surface produced above the Bears Brook volcanics. Maehl (1961, p. 38) found greater irregularity in Pictou County where he measured thicknesses ranging from 50 ft to 240 ft. He reports lithologies similar to those at Arisaig.

LITHOLOGY. A basal conglomerate is present in Doctors Brook (~1.5 ft), on the shore 500 ft southwest of Mill Brook (3 ft), and north of the road to Arisaig Pier (22 ft). It is absent both in McGillivray Brook and on the shore 650 ft southwest of Mill Brook.

The coarse fraction of the conglomerate is variable in size and roundness. Southwest of Mill Brook, it is composed of well-rounded, subspherical pebbles and cobbles as much as 12 cm in diameter. By the pier road at Arisaig and in Doctors Brook, the pebbles are mostly angular. Most of the pebbles are rhyolite and ignimbrite derived from the underlying McGillivray Brook Formation, but some jasper pebbles are present by the pier road, and the Doctors Brook outcrops contain a larger variety of graywacke and igneous pebbles similar to those seen in the Malignant Cove Formation. Many of the larger andesite pebbles have had their outer parts weathered before burial.

Rapid changes in thickness of the conglomerate between Arisaig Point and Mill Brook (Pl. 3C) suggest deposition in channels cutting the surface of the Bears Brook Volcanic Group. The increased variety of pebbles seen in the Doctors Brook outcrops to the east (Pl. 3A) suggests possible correlation with the fluviatile Malignant Cove Formation, but reasons have been given in the previous chapter for considering the Malignant Cove beds to be older than the siltstones of the Beechhill Cove Formation (Fig. 2).

The main part of the Beechhill Cove Formation is composed of very hard quartzose siltstone interbedded with more fissile micaceous shale and a few lenticular bodies of finely cross-laminated sandstone.

The siltstones range in thickness from 0.5 to 18 in. and they make up about 55 percent of the formation. Most are massive, but some are finely laminated–a result of differences in the ratio of quartz to mica. Both types are jointed perpendicular to the bedding, so that they break up into rectangular blocks.

The approximate composition of the siltstone is: 90 percent silt (.01 to .06 mm diam.) consisting of 50 to 60 percent quartz, angular to subangular; 15 to 25 percent chlorite (interstitial, secondary?); 10 percent muscovite, clastic; 5 percent feldspar, altered (including tr. plagioclase, laminated); and tr. opaques (hematite, marcasite or carbon). The remaining 10 percent is clay-sized matrix (note lack of carbonate).

When wet and unweathered, the siltstones are dusky grayish blue-green (5 BG 4/2); when dry and unweathered, they are medium greenish-bluish gray (5 BG 5/1). Weathered surfaces show limonitic staining, which ranges from moderate reddish brown (10 R 5/6) to light brown (5 YR 6/6).

The shale interbeds also range in thickness from 0.5 to 18 in. They make up about 40 percent of the formation. Fifty feet below the top of the formation, the shales become more abundant; in some places, the uppermost 20 ft is all shale.

The approximate composition of the shale is: 80 to 85 percent medium and fine silt (< .04 mm diam.); 30 to 40 percent quartz (angular to subrounded); 25 to 35 percent chlorite (interstitial, secondary?); 10 to 15 percent muscovite, clastic, foliate; 5 percent altered feldspar; tr. opaque (hematite, marcasite, amorphous carbon); 15 to 20 percent clay-sized matrix.

Fresh shale exposures are medium dark gray (N5) when dry, and grayish black (N3) when wet. They weather to medium dark gray (N5) usually with moderate brown (5 YR 5/4) limonite staining.

Sandstones make up less than 5 percent of the formation. These units characteristically have fine cross-laminations and fill shallow channels or lens-shaped scour depressions from 1 to 10 ft across and up to 10 in. deep. The channels are usually cut into the upper surfaces of siltstone beds, rather than into shales. Quite commonly, the basal layer of the channel-filling is fossiliferous, consisting mostly of disarticulated brachiopods. The individual valves are usually whole. The laminations have an average thickness of about 10 mm, and consist of light olive-gray laminae (5 Y 6/1) alternating with dark greenish-gray (5 GY 4/1) laminae. The light-gray laminae are composed of: 95 percent subangular quartz sand averaging 0.067 mm in diameter; 5 percent interstitial clay and sericite.

The darker lamellae contain about: 85 percent quartz sand; 15 percent matrix material.

Measurements of 50 current indicators (cross-laminations of channel sands and a very few sole markings) show that 78 percent of the depositing currents flowed from the northeast quadrant. But it is not known whether these flowed directly off the land mass or whether there were currents parallel to the shore.

STRUCTURE. The beds exposed are less deformed structurally than the higher beds of the Arisaig Group, because they were partially protected from tectonic forces by the more competent volcanics immediately underlying the Beechhill Cove Formation. Not only are they about a mile from the Hollow Fault Zone, but due to the nature of the lithology, deformation of the Bears Brook Volcanic Group has resulted in transcurrent faulting, while the less competent beds above the Beechhill Cove Formation are more usually folded or sheared.

PALEONTOLOGY. Four large fossil collections were made from the Beechhill Cove Formation (*see* Appendix). Collections 10115 and 10117 were made near the base of the formation near Doctors Brook and McGillivray Brook, respectively, and collections 10114 and 10116 were made in the top half of the formation on the southwest side of Beechhill Cove. The brachiopods which make up more than 90 percent of the specimens, have been studied by Harper (1964). No lateral or vertical changes in the faunas were observed. Apart from the brachiopods, no additions to McLearn's (1924) faunal list have been made in the present work.

Harper (1964) and McLearn (1924) both record the following species as being common: *Leptostrophia* [*Rafinesquina*] *beechhillensis* (McLearn), *Mendacella* [*Schizophorella*] *arisaigensis* (McLearn), *Cryptothyrella* [*Whitfieldella?*] *beechhillensis* (McLearn). Lingulids in position of growth and a few horn corals are also present.

AGE AND CORRELATION. Harper (1964, p. 19) states that the above three genera are characteristic of lower Llandovery brachiopod faunas. This age is further suggested by the presence of graptolites at the base of the Ross Brook Formation, which overlies the Beechhill Cove Formation conformably. These graptolites probably belong to the zone of *Monograptus gregarius*, which is partly lower Llandovery and partly middle Llandovery (Berry and Boucot, 1970).

Maehl (1961, p. 78–83) correlates the Beechhill Cove Formation with the upper member of the Glencoe Brook Formation of southeast Pictou County on the basis of *Leptostrophia* [*Eostropheodonta*] *beechhillensis* (McLearn) and *Cryptothyrella beechhillensis* (McLearn) (*see* Harper,

1964, p. 142–144). The Glencoe Brook Formation consists of conglomerates, sandstones, and mudstones–generally much coarser than those of the Beechhill Cove Formation. Its lower member, unfossiliferous, is interpreted to represent a nearer shore facies than the Beechhill Cove Formation.

Ross Brook Formation. The beds of Twenhofel's (1909) Division II were renamed the Ross Brook Formation by Williams (1914, p. 64) after the brook that flows into the sea 800 ft southwest of Arisaig Pier. Dawson (1891) and Ami (1900) called these beds the Arisaig Formation, but Williams (1914, p. 61) used this name for all the marine Silurian and Devonian beds (Arisaïg Group).

Following Maehl (1961), it has been found convenient to divide the formation into three members (although his lower member is different lithologically from that at Arisaig). The relation of these members to the "zones" of McLearn, Williams, and Twenhofel is shown in Table 1.

The lower member is 100 ft thick and corresponds to McLearn's Zone "a" and to Twenhofel's Zone "2." Its lower boundary is well defined in the type section for this member in Beechhill Cove (Pl. 3B), where black shales rest with a sharp contrast on the green shales and siltstones of the Beechhill Cove Formation.

The middle member corresponds to Twenhofel's Zones 3-7 (Table 1). Its base occurs 100 ft above the base of the Ross Brook Formation where the beds become lighter in color due to decrease in carbon; the contrast is gradational over a thickness of 10 ft in Beechhill Cove and Doctors Brook. The upper parts of this middle member contain thin argillaceous siltstones which form about 10 percent of the strata. There is no complete section of this member; the type sections are in Beechhill Cove (for the lower part) and on the shore near the mouth of Arisaig Brook north of Station 110 (for the upper part) (Pls. 3B and 4).

The upper member contains thicker and more abundant siltstones than do the beds below. Its base is at the lowest siltstone bed which is more than 6 in. thick; above this point, siltstones form, on the average, about 50 percent of the strata. On the shore (the type section), the lower boundary of this member occurs about halfway up a 20-ft cliff at Station 110, which is 200 ft southwest of Arisaig Brook (Pl. 4).

DISTRIBUTION AND THICKNESS. The Ross Brook Formation crops out between the mouth of French Brook and the bridge where Nova Scotia Route 45 crosses Doctors Brook (Pl. 1). Throughout this belt, most of the beds lie on the northwest limb of the Arisaig syncline, and have a general southeasterly dip, gentle in the southwest and nearly vertical in the northeast; but there are many minor folds and faults. The formation has a wide outcrop in the Ross Brook syncline, between Arisaig Church and Arisaig Brook upstream from Route 45.

Because of the drift cover, exposures are absent on the southeast limb of the Arisaig syncline, except in Doctors Brook, 1,200 ft southeast of Route 45 (Pl. 5A).

The Doctors Brook syncline has exposures of the upper members on its northwest limb on the cliff north of the west branch of Doctors Brook (at station U48, Pl. 5D). The upper member is also exposed on the southeast limb of this syncline downstream from the junction of the east and west branches (Doctors Brook station M4 to M10, Pl. 5C) and on the valley side 2,000 ft southeast of Route 45 (east of station L21, Pl. 5B).

The thickness of the lower member at Beechhill Cove and Doctors Brook is 100 ft (Pls. 3B, 3A, 5A).

Exposures of the middle member are not good enough to determine the thickness accurately. Twenhofel's estimate of 418 ft for his Zones 3 to 7 on the shore near Arisaig Brook is certainly too low for the whole of the middle member. It does not include the 180 ft of shales seen above the lower member in Beechhill Cove (Pl. 3B). These shales are not interbedded with siltstones (which are present in all except the basal 20 ft of the middle member on the shore at Arisaig Brook) and are therefore lower in the succession than the base of Twenhofel's Zone 3. A study of the map (Pl. 1) would suggest a thickness of 900 ft for the middle member in McGillivray Brook, but signs of faulting are present (just south of Route 45) and some beds may be repeated in this brook. It is concluded that the thickness of the middle member is probably between 600 and 850 ft.

The upper member corresponds to Twenhofel's Zones 8 to 12, for which he gives a thickness of 305 ft. This shore section is not complete, however, as the top of the Ross Brook Formation is faulted against younger beds in Smith Brook and French Brook (Pl. 4). The thickness visible in Arisaig Brook (Pl. 7) 1,300 ft south of Route 4A, is 300 ft, while the McGillivray Brook (Pl. 1) section suggests that 850 ft of the upper member are present. Faulting has perhaps reduced the Arisaig Brook succession and increased the apparent thickness in McGillivray Brook (Pl. 1), but the true thickness cannot be ascertained; it is probably approximately 350 ft.

The total thickness of the Ross Brook Formation at Arisaig is thus around 1,050 to 1,300 ft; but structural deformation is possible, and this estimate can be only very approximate. There may be a 30 percent error.

In Pictou County, Maehl (1961, p. 43) describes 84 ft of mudstone which is correlated with the lower member at Arisaig; he also records

450 ft of mudstone and siltstone in the middle member, and 750 ft of sandstone and mudstone in the upper member; this makes a total of 1,284 ft for the Ross Brook Formation in Pictou County. The total thickness is thus similar to that of the Arisaig beds, but the individual members are different.

LITHOLOGY. The lower member is a carbonaceous feldspathic shale with well-developed bedding-plane fissility. When fresh and dry, the shale is grayish black (N2); it is black (N1) when it is fresh and wet, and it weathers with a moderate brown (5 YR 4/8) limonitic staining.

Thin sections show these beds to be thinly laminated. Single laminations range from 1.0 to 0.06 mm thick, and are differentiated by changes in the ratio of silt to clay. Approximate composition, averaging interlaminar variations, is: 65 percent silt (.05 to .005 mm diam.) and 35 percent clay-sized matrix. The silt fraction consists of 20 to 30 percent quartz, subangular; 15 to 20 percent chlorite, interstitial, secondary; 10 to 15 percent muscovite, clastic; 5 to 10 percent opaques, marcasite, carbon and/or iron oxide; and 5 percent altered feldspar.

At least 13 layers of tuff, ranging in thickness from 0.5 to 4.0 in. are interbedded with the shales in the lower half of the lower member. They are composed of mica, calcite, and iron oxides 0.25 to 0.02 mm in diameter, imbedded in a mottled matrix of silt-sized feldspar, sericite, and secondary carbonate. The coarser clasts are concentrated toward the base, and the average grain size decreases toward the top of each tuff layer. Some of the tuffs show shear zones oriented oblique to the bedding and filled with secondary calcite. Two 2-in. beds, 35 and 40 ft from the base, contain brachiopod shells that have served as nuclei for the precipitation of calcareous nodules, which range up to 1 ft in diameter and 3 inches in height.

The middle member of the Ross Brook Formation is composed of fissile shale, with some interbeds of siltstone in the higher parts.

The shale is composed of: 80 percent silt (.05 to .06 mm diameter) and 20 percent clay-sized matrix. The silt fraction consists of 25 to 35 percent muscovite, clastic, foliated; 15 to 25 percent quartz, subangular; 10 to 20 percent chlorite, interstitial, secondary; 5 to 10 percent opaques, marcasite, iron oxide and/or carbon; and 5 percent altered feldspar.

The shale of the middle member is dark gray (N3) when fresh and dry, grayish black (N2) when wet, and light brown (5 YR 5/6) when weathered and stained with limonite. Some shale beds are microscopically laminated (up to 140 laminae per centimeter); the laminae have differing ratios of silt to clay.

The siltstones are absent in the basal 180 ft of the middle member exposed in Beechhill Cove (Pl. 3B). They first appear in the beds on the shore south of Arisaig Pier at a point 220 ft northeast of station 126 (Pl. 4). Beyond this point, they comprise about 10 percent of the strata, and range in thickness between 0.5 and 5.0 in. The siltstones are microscopically laminated and rarely cross-laminated—a consequence of rhythmic alternations in the mica content (higher in the darker laminae). Most of the particles lie within the coarse or medium silt fraction (0.06 to 0.15 mm).

The approximate composition is: 55 percent angular and subangular quartz; 5 percent sericitized feldspar; 5 percent muscovite (oriented parallel to bedding); and 35 percent overgrowths of chlorite and marcasite.

The alternations between blocky siltstones and fissile shale beds is explained by the variations in muscovite content and alignment of the flakes parallel to the bedding.

The lower part of the middle member is not fossiliferous, and worm-disturbed beds are rare. Southwest of shore station 122 (Pl. 4), fossils, trails, and burrows are more abundant in the higher parts of the member.

In the upper member of the Ross Brook Formation, the ratio of siltstone to shale averages about one to one, in contrast to the one to ten ratio in the middle member. The shales and siltstones of the upper member are similar in color, composition, and grain size to those of the middle member. However, no laminations such as those in the shale of the middle member were detected. Some of the thin sections show cylindrical variations in texture oriented normal to the bedding. That they are probably burrows is substantiated by the occurrence of infaunal bivalves in living position in their burrows. Bedding planes in the shales commonly show tracks and trails.

Fossils are rare in many of the siltstones. Most of the siltstones lack laminations and cross-bedding, but some show both these structures, together with oscillation ripples near their bases. Larger ripples (1 ft wavelength, 2 in. amplitude) may have brachiopod shells in their depressions (for example, at shore station 105, Pl. 4). The bases of the siltstone beds frequently show load casts (well displayed at station 107, where they are 4 in. wide and 2 in. deep), and sometimes scours up to 4 inches depth (for example, station 102). The shore section also shows (Pl. 4) some graded siltstones in Twenhofel's Zone 10 (station 95). Cross-bedding and brachiopods occur in the same beds, and so it seems probable that these graded beds were deposited in the same shallow-water environment as was the rest of the upper member.

A third lithological type, arenaceous limestone or calcarenite, is present in the upper member; it makes up less than 5 percent of the strata. It has grains ranging from 0.5 to 0.02 mm and is usually laminated. The laminae are between 0.1 and 4.0 mm thick, which result from alternations in the proportions of quartz and mica. Quite commonly, the basal lamination, thicker than the others, contains a high concentration of disarticulated bivalve and brachiopod shells lying convex upward.

The approximate composition of the calcarenites is: 55 percent carbonate (interstitial and probably derived from included shells); 25 percent subangular quartz; 10 percent chlorite; 5 percent feldspar; 5 percent mica; and trace carbon.

Because of the lack of cross-lamination, current ripples, sole markings, channels or other indications of current directions, no statistical studies of flow directions of depositing currents were made.

STRUCTURE. Shales form over 90 percent of the beds in the lower and middle members of the Ross Brook Formation and about 50 percent of the upper member. The formation is thus very incompetent. The irregularities in the major folds that are expressed by transcurrent faults in the Bears Brook Volcanic Group and the Beechhill Cove Formation take the form of numerous small folds and faults in the Ross Brook Formation. These are especially numerous in the middle and upper members, which, although they contain siltstones, are both more distant from the competent underlying rocks and closer to the disturbed region near the Hollow Fault.

PALEONTOLOGY. The basal 15 ft of the lower member of the Ross Brook Formation contains graptolites. These were first recorded by O. T. Jones (1926) from the south end of Beechhill Cove and from Doctors Brook (a few yards above the old bridge near Route 45). The assemblage collected from Beechhill Cove (Pl. 3B, locality 10119) has been identified by W.B.N. Berry; it contains *Climacograptus scalaris* var. *normalis* Lapworth, *Climacograptus rectangularis* M'Coy, *Monograptus acinaces* Tornquist, *Monograptus lobiferus* M'Coy, and *Orthograptus bellulus* Tornquist (Berry and Boucot, 1970).

The brachiopods occurring in nodules in the tuff bands 35 ft above the base of the lower member (locality 10118) in Beechhill Cove (Pl. 3B) are identified as *Pentlandina* sp.

No fossils have been found in the upper part of the lower member or the lower part of the middle member. Worm-disturbed beds appear 50 ft northeast of station 122 (Pl. 4) on the shore south of Arisaig Pier; the lowest fossil bed (locality 10121) is at station 122, where (among other brachiopods and also pelecypods) Ziegler has identified *Eocoelia (Coelospira) hemisphaerica* (J. de C. Sowerby). A stratigraphically higher collection (10122) 10 ft southwest of station 117 (Pl. 4) has yielded *Eocoelia curtisi* Ziegler. Ziegler's (1966) study of *Eocoelia* would strongly suggest that Hall's (1860) type of *E. intermedia* came from a bed between localities 10121 and 10122, as it is intermediate in its evolutionary development between *E. hemisphaerica* and *E. curtisi.*

Graptolites are present 50 ft southeast of station 112 (Pl. 4), just east of the mouth of Arisaig Brook (locality 10123); the collection (identified by Berry) consists of *Monograptus* cf. *marri* Perner, and *Retiolites geinitzianus* var. *angustidens* Elles and Wood. Brachiopods and bivalves also occur in many beds near the mouth of Arisaig Brook, but the only other large collection made in this area (10124) (Pl. 7A) yielded no fossils of great stratigraphical value.

Brachiopods, epifaunal species of bivalves, and graptolites occur in many beds throughout the upper member of the Ross Brook Formation (*see* Appendix). Four collections (10126–9) were made on the shore (Pl. 4) at well-spaced intervals; three yielded *Eocoelia sulcata* (Prouty) (10127–9; identifications by Ziegler) and two (10128–9) contained *Monograptus clintonensis* (Hall) (identified by Berry). Other collections from the upper member include *Eocoelia sulcata* (Prouty) from Doctors Brook (10130) (Pl. 5A) and Arisaig Cemetery (10131) (Pl. 2).

The faunas can all be assigned to the shallow-water *Eocoelia* Community (Ziegler and others, 1968).

AGE AND CORRELATION. Berry and Boucot (1970) assign the graptolite assemblage from the lower member (10119) to the zone of *Monograptus gregarius*, which they place astride the boundary between the lower and middle Llandovery. This age is significantly older than the age given by Jones (1926) who assigned these beds to the base of the upper Llandovery, but it agrees well with the brachiopod ages.

The fossiliferous part of the middle member includes both *Eocoelia hemisphaerica* (of C_1 to C_2 age) and *E. curtisi* (C_5). This member thus certainly represents the first five of the six brachiopod zones of the upper Llandovery (Williams, 1951; Ziegler, 1966). The unfossiliferous lower part of this middle member is probably older than C_1, and is tentatively assigned to the middle Llandovery on the grounds that no break in sedimentation has been observed between localities 10119 and 10121.

The graptolites in the upper member of the Ross Brook Formation indicate a late Llandovery age. However, *Eocoelia sulcata* is confined to C_6 and the lower Wenlock; moreover, Ziegler is of the opinion that the forms present throughout the upper member have ribs indicative of C_6 age (the Wenlock forms are quite smooth). It is therefore concluded that all of this member is C_6 in age.

French River Formation. Maehl (1961, p. 51) proposed this term for arenaceous beds conformably overlying the Ross Brook Formation on the French River in Pictou County at a point 1,050 ft south of the Nova Scotia Route 4 bridge, 10 mi east of New Glasgow. At that time, it was thought that these beds were not represented at Arisaig, and Maehl considered the French River Formation to be older than the McAdam Formation as defined by Williams and McLearn, although no contact between those two formations was seen in Pictou County. However, subsequent fossil collecting has shown that species characteristic of the French River Formation occur at Arisaig in the lower part of the McAdam Formation of earlier workers (Zone a of McLearn; Zone 5 of Williams). Moreover, the only differences in lithology between those beds at Arisaig and those on the French River have been induced tectonically. The beds at Arisaig are more altered and sheared than their equivalents in Pictou County. In view of this discovery, and in view of the inclusive and diverse nature of the old McAdam Formation, it is now considered desirable to rename these beds as shown in Table 1 (*see also* Table 3).

The extension of the French River Formation to the Arisaig area allows a slightly clearer definition of the formation. Top and bottom contacts are exposed along Arisaig Brook (Pl. 7), 50 ft upstream and 70 ft downstream from the 2½-ft iron-ore bed, respectively (although, as the formation is thicker in other brooks, some strike faulting is suspected here–*see* discussion on thickness below). The basal contact is exposed only in Arisaig Brook (elsewhere in the Arisaig area, it is faulted or covered); although the contact is conformable, there is a sharp change in lithology from the dark shales and siltstones of the Ross Brook Formation to the lighter siltstones of the French River Formation. The contact with the overlying Doctors Brook Formation is gradational, and the boundary is taken where shales become the dominant lithological type. In the French River Formation 85 percent of the beds are siltstones; but in the adjacent formations, the percentage is much lower (50 percent in the upper member of the Ross Brook Formation; 35 percent in the Doctors Brook Formation).

In both the Arisaig area and Pictou County, the following brachiopods are confined to the French River Formation: *Heterorthella maehli* Harper, Boucot, and Walmsley, *Meristina billingsi* (Dawson), *Plagiorhynchia* cf. *plastica* (McLearn), *Striispirifer stonehousensis* (McLearn). These diagnostic fossils were the principal means of first identifying the French River Formation in structurally complex areas such as the Doctors Brook Valley (Pl. 5).

DISTRIBUTION AND THICKNESS. The French River Formation is not exposed on the shore, but it does occur only 100 ft from the mouth of

French Brook and 170 ft above the waterfall where Smith Brook enters the sea (Pl. 4). In both brooks, the upper and lower contacts are faulted.

The formation is seen along the northwest limb of the Arisaig syncline in seven brook sections from Arisaig Brook to Beechhill Cove Brook (Pl. 1). Exposures are absent on the southeast limb of this syncline because of the drift cover. The French River Formation is exposed for the greater part of 4,000 ft along Doctors Brook (Pl. 5) between station U42 (on the west branch) to station L12 (a point 1,200 ft south of Nova Scotia Route 45). The beds lie in the Doctors Brook syncline, but have a very complex structure of small folds and faults due to their proximity to the Hollow Fault.

Complete sections of the French River Formation are exposed in three brooks on the north limb of the Arisaig syncline. Apparent thicknesses (in feet) are as follows:

	Arisaig Brook (Pl. 7)	Ross Brook South Branch (Pl. 6)	McGillivray Brook (Pl. 1)
Total thickness	123	240+	280
Thickness above iron ore	50	125	iron ore absent

From these figures, it would appear that 280 ft is the best estimate. It seems that more than one-half the formation is cut out by faulting in Arisaig Brook; the position of the iron-ore bed shows that more than one fault is involved, this may also be true of the sections in the other brooks and errors in these estimates may be as high as 30 percent.

The iron-ore bed was formerly mined in Arisaig Brook (Pl. 7) and the south branch of Ross Brook (Pl. 6). It also may have been worked in the gently dipping beds between stations 98 and 102 on the north branch of Ross Brook, where considerable iron-ore debris (some blocks up to 3 ft) occurs. The ore bed is absent east of Ross Brook.

In Pictou County, the type section of the French River Formation shows 175 ft of beds, but no upper contact is seen. Some beds formerly ascribed to the lower member of the McAdam Formation are also present in northeast Pictou County; Dawson (1881) and Lindeman and Bolton (1917) indicate the presence of the iron-ore bed, but Maehl (1961, p. 55) was unable to find it. However, most of Maehl's "lower McAdam" is equivalent to our Doctors Brook Formation.

In the Lochaber area of southwest Antigonish County, D. Benson (Canada Geol. Survey) has found *Meristina billingsi* 1,200 ft upstream from the mouth of the brook 6,500 ft north of McNab Brook on the west side of Lake Lochaber (B079 B2963); this typical French River brachiopod is the oldest known from the Silurian of this area.

LITHOLOGY. Approximately 85 percent of the French River Formation consists of blocky beds of siltstone (1 to 12 in. thick). They are interbedded with shale that makes up most of the remainder of the formation. The fissility of the shales is commonly interrupted by a rough fracture cleavage. A 2½-ft bed of fossiliferous hematite is present in Arisaig Brook and Ross Brook.

Most of the massive siltstones are fine grained and poorly sorted in sizes less than 0.1 mm. Fresh exposures range between light gray (N7) and medium bluish gray (5B 5/1); the bluer units are more argillaceous. The siltstones weather to light olive gray (5Y 5/1) or light, limonitic brown (5 YR 5/5). Composition varies between beds, but the average of the 60 percent coarse silt (0.06 mm average diameter) is approximately as follows: 40 to 50 percent quartz, angular, well sorted; 5 percent muscovite, clastic; 5 to 10 percent opaque, iron oxides, marcasite and/or carbon. The 40 percent fine silt matrix consists of 25 to 30 percent sericite, interstitial, secondary; 10 to 15 percent chlorite (note lack of carbonate).

Irregular variations in the silt/clay ratio result in some siltstones being slightly fissile. These beds were probably disturbed by burrowers. Others show graded bedding and cross-laminations similar to the siltstones of the upper member of the Ross Brook Formation (Pls. 5A, 5B, Doctors Brook Stations L16 and L28).

The shale beds range in thickness from 1 to 7 in. They are medium dark gray (N4) when fresh, but weather to grayish black (N5); both limonite and black manganese staining are common. The shales are fissile, a result of alignment of mica which is quite evident in thin section. The shales also show microscopic crude laminations, which, like the sandstones, seem to have been stirred by burrowers.

Composition varies from bed to bed, especially in the proportions of fine sand, silt, and clay. The finer units have about 40 percent medium silt (−.025 mm avg diam.) consisting of 35 percent quartz, subrounded; 5 percent opaques, marcasite, iron oxides and/or carbon; and 60 percent fine silt consisting of 40 percent sericite, secondary, interstitial.

The dark reddish-brown (10R 3/3) iron-ore bed contains 90 percent finely oölitic hematite, and 10 percent well-sorted, subrounded to well-rounded, medium-fine quartz sand (averaging about 0.25 mm). This quartz is coarser and more rounded than that in most other Arisaig rocks. Some shell lenses are present with brachiopods, crinoids, and orthoceratids in various stages of fragmentation. Brachiopods, especially *Meristina billingsi* (Dawson), are abundant in the iron ore; similar forms are present in the surrounding sandstones. Trails on bedding planes have been observed on Doctors Brook at station L38, and ripple marks at station L27 (Pl. 5B).

The sedimentary structures in Doctors Brook have been examined more closely than elsewhere, mainly in order to determine upward direction criteria.

No statistical studies were made of current direction because of a lack of channeling, cross-lamination, sole markings, and other evidences of flow.

STRUCTURE. The beds of the French River Formation on the northwest limb of the Arisaig syncline are steeply dipping, faulted, and sheared, but probably the least disturbed section is in McGillivray Brook. The numerous minor folds in Doctors Brook (Pl. 5) are described in the general chapter on structure. Faulting has produced three outcrops of the iron-ore bed in Ross Brook, south branch (Pl. 6).

PALEONTOLOGY. Brachiopods are common throughout the French River Formation (*see* Harper, 1964); and the following species are confined to this formation: *Heterorthella maehli* Harper, Boucot, and Walmsley, *Meristina billingsi* (Dawson), *Plagiorhynchia* cf. *plastica* (McLearn), and *Striispirifer stonehousensis* (McLearn). Graptolites, assigned to *Monograptus* cf. *distans* (Portlock) by W.B.N. Berry, have been found at locality 10133 (25 ft southwest of station M25 in Doctors Brook (Pl. 5B). Bivalves, trilobites, orthoceratids, and crinoids are also present in some beds (*see* Appendix).

Age and Correlation

Monograptus cf. *distans* belongs to a group with hooked thecae and rather slender rhabdosomes that occurs most frequently and in greatest diversity in the late Llandovery, but some range up into the Wenlock (W.B.N. Berry, 1965, personal commun.).

Of the brachiopods, most are long-ranging species, but *Plagiorhynchia* is known elsewhere only from Llandovery- and Wenlock-age strata (Harper, 1964, p. 23).

The French River Formation is conformable on the upper member of the Ross Brook Formation (which is known to be C_6). A lower Wenlock age thus seems certain for the greater part of the formation; the basal part could, however, be upper Llandovery.

Doctors Brook Formation. This term, used informally by Honeyman (1870), is here employed as an equivalent to Twenhofel's (1909) Zones 13 to 20 and to McLearn's (1924) Zone b of his "McAdam Formation" (Table 1). It corresponds to most of Maehl's (1961) "lower McAdam Formation." However, Maehl, following Williams (1914), included Twenhofel's Zone 21 with these beds; Zone 21 consists of black shales and is now included, with similar shales higher in the sequence, in the overlying McAdam Brook Formation.

It is estimated that some 25 ft of beds lower than Twenhofel's Zones

13 through 20 (that is, lower than any of this formation exposed on the shore) are exposed in inland sections, such as Doctors Brook, west branch (Pl. 5) and Ross Brook, south branch (Pl. 6) (*see* discussion on thickness below). In these brooks, the bottom contact is gradational over about 10 ft, the Doctors Brook Formation being distinguished from the underlying French River Formation by the presence of numerous calcareous beds (siltstone and limestone) and by the increase in the proportion of shale beds to more than 50 percent of the strata.

On the shore, the upper contact of the Doctors Brook Formation is sharp; it is well exposed (at stations 51 and 52, Pl. 4) northeast of McAdam Brook, where black shales (of the McAdam Brook Formation) rest on gray shales and calcareous siltstones. At this locality, the top ⅛ in. of the Doctors Brook Formation contains scours or borings filled with black silt. In contrast, this top contact is gradational in Arisaig Brook (Pl. 7) and in Ross Brook (Pl. 6). The boundary is placed where there is a reduction in the number of 2-in. siltstone beds and the shales become darker.

The type section of this formation is taken on the shore northeast of McAdam Brook, between station 51 and the lowest outcrop northwest of station 59 (Pl. 4).

DISTRIBUTION AND THICKNESS. The Doctors Brook Formation is well exposed on the shore between French Brook and McAdam Brook. Near French Brook (at station 83) (Pl. 4), the beds are faulted down against the Ross Brook Formation, and neither the top nor the base of the formation is exposed. To the southwest of station 83, four other important faults (there are numerous smaller faults) trending between north and northwest have thrown down the northeast blocks and have caused a repetition of beds. An almost complete section is present from the fault near station 58 southwest to the contact with the overlying McAdam Brook Formation. Because of the faulting, the present estimate of the thickness (275 ft) is much lower than that given by Twenhofel (544 ft).

Inland, the formation is exposed on the northwest limb of the Arisaig syncline between Arisaig Brook and Beechhill Cove Brook (Pl. 1). But only the Arisaig Brook and Ross Brook sections show more than the lower half of the formation.

In Doctors Brook, numerous exposures of all parts of the formation are present in the west and east branches between stations U35 and U82 (Pls. 5C and 5D). The basal part of the formation is also seen between stations M11 and M14 (Pls. 5B and 5C) in the axial region of the Doctors Brook syncline. The lower contact of the Doctors Brook Formation is exposed in the west branch (between stations U48 and MO, Pl. 5D), but the upper contact is not exposed because of a 100-ft covered interval between stations U82 to U84 (Pl. 5C).

Apparent thicknesses exposed are as follows (in feet):

SHORE (PL. 4, STATIONS 51 TO 58)	ARISAIG BROOK (PL. 7)	ROSS BROOK SOUTH BRANCH (PL. 6)	DOCTORS BROOK WEST BRANCH (PL. 5)
250+ (base not seen)	500	275	? 500

The beds in Ross Brook are less sheared and deformed than those in Arisaig Brook and Doctors Brook, and it is considered probable that faulting has repeated parts of the succession in these latter sections; one such fault is exposed in Arisaig Brook 120 ft downstream from station 81 (Pl. 7). The thickness of 275 ft in Ross Brook is therefore taken as the best estimate. Errors up to 30 percent may be present due to structural deformation but, if this estimate is correct, it follows that the shore section between stations 51 and 58 is almost complete; this section shows 250 ft of beds below the top of the formation, so only the basal 25 ft of the Doctors Brook Formation may be covered at low tide.

LITHOLOGY. The approximate composition of the Doctors Brook Formation is: 55 percent shale, 35 percent thin-bedded calcareous siltstones and silty limestones, and 10 percent shelly limestones (or coquinites).

The composition of the shale averages about: 40 percent clastic silt (.03 mm avg diam) consisting of 20 percent quartz, subangular; 10 percent altered feldspar; 5 percent muscovite, clastic; 5 percent opaques, marcasite, iron oxide and/or carbon; and 60 percent matrix ($\leq$.015 mm avg diam.) consisting of 40 percent muscovite, interstitial, foliate; 15 percent chlorite, interstitial, secondary; and 5 percent calcite, interstitial, secondary.

The shale is medium gray (N5) when fresh, but weathers to medium dark gray (N4). Some of the shale beds are laminated; the lamellae, ranging in thickness from 0.25 to 1.0 mm, result from alternations in the ratio of quartz to mica.

At places, the shales contain dark phosphatic inclusions, some of which are cored by inarticulate brachiopods, but most of which appear inorganic. These phosphatic nodules are well developed on the shore at stations 52, 55, and 79 (Pl. 4) and on the west branch of Doctors Brook at station U48 (Pl. 5D); 150 ft along strike (Doctors Brook, station U50, Pl. 5D) bedding-plane trails are present in these beds.

About 35 percent of the sediments are more indurated and resistant than the shales. They occur in lenses, beds, and channel fillings. Most of them range in thickness from 2 to 5 in., and they are generally much thinner than the coarse siltstones of the French River Formation. They range in composition from calcareous siltstones to quartzose limestones. These beds are generally laminated or cross-laminated, and their darker lamellae contain a higher proportion of detrital mica. A few beds show grading. The best exposed channel extends east-southeast on the shore midway between stations 76 and 77 (Pl. 4). It is 6 in. wide and 6 in. deep, is filled with fine clastic quartz, and cuts into a shale bed. In 1964, 3 ft of its length was exposed.

The more quartzose units are medium gray (N4) and average about 60 percent medium and coarse silt (.05 to .005 mm diam.) consisting of 35 to 45 percent quartz, rounded; 5 to 10 percent muscovite clasts; 5 percent altered feldspar; 5 percent opaques (marcasite, iron oxides and/or carbon); and 40 percent matrix ($<$.02 mm diam.) consisting of 25 percent sericite, interstitial, secondary; 10 percent chlorite; and 5 percent calcite.

The quartzose limestones or coquinites, which can be as much as 9 in. thick, are bluish gray (5B 6/1), and average about: 65 percent carbonate; 25 percent angular quartz; 10 percent chlorite, interstitial, secondary; 5 percent feldspar; trace muscovite; and trace chlorite.

These coquinites, comprising less than 10 percent of the formation, usually occur in lenses or channel deposits scoured into underlying shale. They are composed of calcite-filled shell molds, imbedded in a matrix of silt. Many of these beds grade upward into laminated or cross-laminated siltstones. Coquinite lenses 6 in. thick, composed mainly of brachiopods, are well exposed on the shore near stations 61 and 81 (Pl. 4).

Insufficient data were collected from the Doctors Brook Formation to make a reliable assessment of current indicators, but 11 out of 18 measured cross-beds show depositing currents from the northeast quadrant.

STRUCTURE. The shore outcrops between McAdam Brook and French Brook show gentle dips and comparatively little disturbance of the beds (Pl. 4). Although minor faults are present, accommodation to stress has been largely along five fault planes. The rocks generally show little sign of shearing or other deformation, although cleavage is developed to the northeast of the fault at station 75. Pseudo-ripples, due to oblique intersection of bedding and cleavage, are present between stations 76 and 77 (Pl. 4). The most complete, and also the least disturbed, section on the shore is between stations 51 and 59 (Pl. 4).

On the northwest limb of the Arisaig syncline between Arisaig Brook and Beechhill Cove Brook, the beds are faulted and sheared. Of the two complete sections on this limb, in Arisaig Brook and Ross Brook, the latter is the less deformed.

The Doctors Brook syncline, best exposed along Doctors Brook, is obscured by complex minor structures that are described in the general chapter on structure.

PALEONTOLOGY. Although fossils are much more abundant in this formation than in the underlying French River Formation, the same species of brachiopods are common to both (*see* Appendix). The following species are confined to these two formations: *Sphaerirhynchia* sp. [*Camarotoechia obtusiplicata* McLearn *non* Hall], *Isorthis macadamensis* Harper MS, *Idiorthis matura* (McLearn), and *Amphistrophia funiculata* (M'Coy).

R. K. Bambach (1967, personal commun.) states that both epifaunal and infaunal bivalves are present, but it is not yet known whether these assemblages are identical in both Doctors Brook and French River Formations.

Age and Correlation

The Doctors Brook Formation lies immediately below beds containing the lower Ludlow graptolite, *Monograptus varians* Wood; it occurs above beds with *M.* aff. *distans* (Portlock) to which W.B.N. Berry has assigned a late Llandovery or early Wenlock age.

In the previous section, reasons were advanced for assigning a lower Wenlock age to the underlying French River Formation. The Doctors Brook Formation is older than the lower Ludlow; therefore, it is probably of upper Wenlock age.

McAdam Brook Formation. The McAdam Brook Formation is a new term used in a different sense from the McAdam Formation of earlier authors. As now defined (Table 1), it is equivalent to Twenhofel's (1909) Zones 21 to 27, to McLearn's (1924) Zone c, approximately equivalent to Zone 8 of Williams (1914), and to Maehl's (1961) upper member of the McAdam Formation (but these latter two authors included Twenhofel's Zone 21 with the underlying beds). The base of the formation is exposed 200 ft northeast of the mouth of McAdam Brook (Pl. 4, between shore Stations 51 and 52) where black micaceous shales rest on the gray shales and siltstones of the underlying Doctors Brook Formation.

The formation is divided into two members. The lower member consists of 150 ft of black shales and siltstones with calcareous nodules; it crops out between a point 30 ft southeast of McAdam Brook station L25 (Pl. 8) and Shore Station 51 (Pl. 4). It corresponds in lithology to Twenhofel's Zones 21 to 25, but his thicknesses for these zones total 404 ft due to repetition of beds along the shore southwest of McAdam Brook (Twenhofel's description refers to the shore section).

The upper member consists of 140 ft of thicker siltstone beds (many over 2 in.) and shales exposed along McAdam Brook (Pl. 8) between

30 ft southeast of station L25 and station L22. Lithologically they correspond to Twenhofel's Zones 26 and 27. Twenhofel's thickness for these zones is only 71 feet. However, the beds on the shore, which Twenhofel measured, are considerably disturbed. On the other hand, along this part of McAdam Brook, there is only one small fault (between stations L24 and L25, Pl. 8), consequently, the McAdam Brook exposures are therefore adopted as the type section of the formation.

The upper contact is gradational; the siltstones become thicker and greener near the top of the upper member, and where they become the dominant lithology, with many beds more than a foot thick, the sediments are assigned to the overlying Moydart Formation. This upper contact crops out in McAdam Brook at station L22 (Pl. 8), and on the shore between stations 37 and 38 (Pl. 4).

DISTRIBUTION AND THICKNESS. The McAdam Brook Formation is well exposed on the shore (Pl. 4) between stations 37 and 52 (a distance of 1,300 ft), and in the adjacent parts of McAdam Brook (described above). It is also exposed along Smith Brook and French Brook (Pl. 4), but at these places only the lower member is exposed, and the upper and lower contacts are faulted out.

Inland, the only complete section of this formation is along Arisaig Brook (Pl. 7), but the lower member can be traced northeastward along the northwest limb of the Doctors Brook syncline, as far as Ross Brook (Pl. 6). All the contacts along Arisaig Brook are gradational.

Along Doctors Brook (Pl. 5), beds of similar lithology to that of the McAdam Brook Formation are seen on the west branch between stations U6 and U18 and at U23 and on the east branch between stations U83 and U87, and at U96 and U97. All of these outcrops are very close to the Hollow Fault; consequently, the beds are sheared and structurally disturbed and no recognizable fossils have been found. The diagnostic lithology of the lower member—calcareous nodules in black shales—appears between stations U6 and U7 (Pl. 5).

The apparent thicknesses are as follows (in feet):

	McADAM BROOK (PL. 8)	ARISAIG BROOK (PL. 7)
upper member	140	120
lower member	150	260

Because of numerous small faults and folds and a lack of marker beds, it is not possible to make any assessment of the thickness of the McAdam Brook Formation on the shore (*see* remarks above), in Doctors Brook, or in Arisaig Brook. The best estimate of thickness is the 290 ft obtained from the McAdam Brook section, where only slight folding and faulting appears to be present, but even here the error may be as much as 30 percent.

The McAdam Brook Formation is probably present west of Lochaber Lake.

LITHOLOGY. The basal 50 ft of the lower member of the McAdam Brook Formation consists of very fissile grayish black (N2) papery silty shale with less than 10 percent lighter, quartzitic interbeds less than 2 in. thick. Both the black shale and the quartzose interbeds have fine bedding plane fissility.

The shale is composed of: 80 percent fine silt (.02 mm mean diam) made up of 30 percent quartz, subangular; 20 percent sericite; 10 percent chlorite, secondary interstitial; 10 percent feldspar; 5 percent calcite; 5 percent opaques (marcasite, iron oxides and/or carbon). The remaining 20 percent of the particles are of clay size and probably of mineralogy similar to the silt material.

The upper two-thirds of the lower member is 85 percent fissile dark gray (N3) shale. The remaining 15 percent consists of thin blocky interbeds of siltstone and arenaceous limestone. Septarian nodules are also characteristic.

The siltstones range between 0.5 and 4.0 inches in thickness. Fresh exposures are medium dark gray (N4), but the rock weathers with a grayish-orange (10 YR 7/4) limonitic stain. They are finely laminated and cross-laminated, the laminae ranging between 1.5 and 0.1 mm thick. The darker laminae have a higher proportion of mica.

The arenaceous limestones in the upper two-thirds of the lower member are medium dark gray (N4), and occur in beds 0.5 to 3.0 in. thick. They are either massive or have microscopic laminations, 0.02 to 1.0 mm thick. The darker laminae are much richer in clay-sized matrix materials. Some lenticular bodies of shelly limestone 0.5 to 3.0 in. thick and up to 4 ft in diameter were deposited in depressions scoured in underlying muddy sediments. Although the mineralogy is quite variable between beds, the average composition of these calcareous rocks is about: 45 percent carbonate; 20 percent quartz, subangular; 15 percent chlorite; 10 percent sericite; 5 percent altered feldspar; and 5 percent opaques (marcasite, iron oxide, and/or carbon). The clastic sediments, included in the limestone, are silt-sized particles.

Septarian nodules occur on the shore between stations 41 and 49 (Pl. 4); they are also present in McAdam Brook (stations L25 to L27, Pl. 8),

Arisaig Brook (stations 85 to 86, Pl. 7), and Doctors Brook (stations U6 to U7, Pl. 5). The septarian nodules have been precipitated around nuclei (probably shells) in the laminated siltstones. They often have an outer siltstone crust 2 or 3 in. thick and a core of calcite crystals. The outer crust is pierced by wedge-shaped cracks, filled with calcite, which taper inward toward the center of the nodule. The nodules average about 8 inches in thickness and 18 inches in diameter; the largest one found measures 24 by 36 in. Some nodules show cone-in-cone structures along their margins, and cone-in-cone structure is also present in calcareous beds where there are no nodules.

The upper member of the McAdam Brook Formation is transitional between the lower member and the overlying Moydart Formation. The lithology shows an upward gradation from gray shales with interbedded siltstones (usually more than 2 in. thick, but otherwise similar to the siltstones in the lower member) into green siltstones (up to 12 in. thick) with subordinate shales. Large load casts occur on the shore cliffs at station 38 (Pl. 4), but they are rapidly being destroyed by wave erosion.

Apart from lenticular fossil-filled scours, no clear evidence of channeling has been found in the McAdam Brook Formation. Thirty-two current directions have been determined from cross-laminations and, in a few cases, current ripples; 75 percent of these indications show currents flowing from the northeast quadrant (Fig. 14).

STRUCTURE. Except in McAdam Brook, the predominantly incompetent beds of the McAdam Brook Formation show many minor folds and faults. Near the mouth of McAdam Brook, many of the black shales of the lower member show irregular fracture cleavage. On the shore, between stations 42 and 45, low-dipping cleavage is strongly developed (Pl. 4).

In most exposures, only a general impression can be made as to the amount of deformation of the incompetent rocks, but at shore station 45 (Pl. 4), deformed rhynchonelloid brachiopods occur in pencil shales. The shells are all compressed to three-quarters of their original dimensions in a northeast-southwest direction, and are similarly stretched in a northwest-southeast plane. The pencil shales are elongate horizontally east and west, parallel to the strike of the south-dipping beds.

PALEONTOLOGY. The basal 50 ft of shales of the lower member of the McAdam Brook Formation have yielded *Monograptus varians* Wood from locality 10168, which is 45 ft south of shore station 50 (Pl. 4). The specimens were identified by W.B.N. Berry.

Brachiopods are common throughout all of the formation (*see* Appendix); two large collections (10169, 10170) show a dominance of

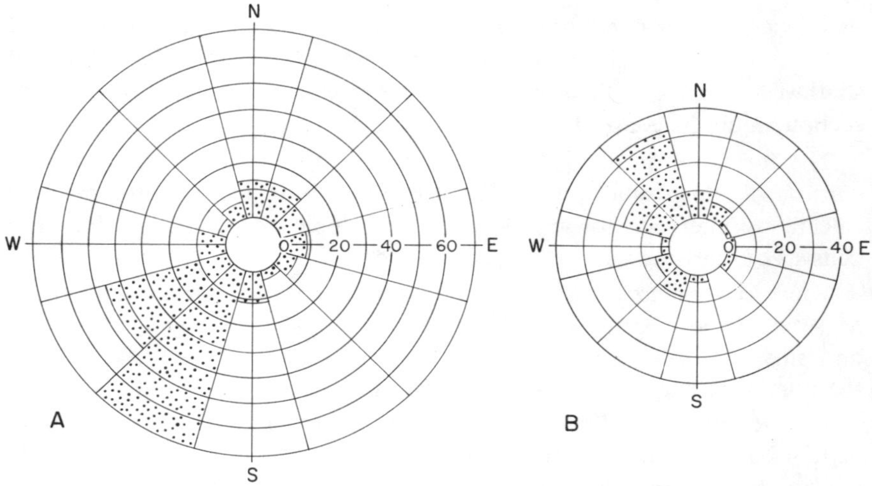

Figure 14. Current directions: diagram showing directions toward which currents flowed. A. 202 current indicators from all formations Beechhill Cove through middle Stonehouse (l. Llandovery to m. Downtonian) showing dominant flow toward the southwest. B. 96 current indicators from upper Stonehouse and Knoydart Formations (upper Downtonian and Dittonian) showing dominant flow toward the northwest (*see also* Table 6).

rhynchonelloids, although *Salopina, Protochonetes,* and *Howellella* are also common (Harper, 1964). The only brachiopod species confined to this formation is *Plagiorhynchia* sp. A, Harper MS. R. K. Bambach reports that the bivalves are mostly infaunal species, although several concentrations of epifaunal forms occur, they are particularly abundant between shore stations 41 and 46 (Bambach's localities A3612 to A3641, Pl. 4). A few trilobites are recorded by McLearn (1924) and Twenhofel (1909).

AGE AND CORRELATION. W.B.N. Berry assigns an early Ludlow age to the graptolites at locality 10168; this is in stratigraphical agreement with McLearn's (1924, p. 35-36) record of *Monograptus wandalensis* Watney and Welch "from near the top of Zone c of the McAdam Formation." Thus the age of the lower member of the McAdam Brook Formation is established. The specimens recorded by Twenhofel (1909) from his zone 24 have been reidentified by Berry as *Monograptus fritschi* Perner; they are probably the same specimens as those examined by McLearn; Berry assigns them to the lower Ludlow zones of *M. scanicus/tumescens.*

Moydart Formation. Ami (1900) named this formation but did not define it precisely. It corresponds to Twenhofel's (1909) zones 28 to 32, to Williams' (1914) zones 9 and 10 (Table 1), and is used here in the same sense as by these authors and as by McLearn (1924). It should be noted that Twenhofel inadvertently omitted zone 29 in his 1909 paper; this zone is 97 ft thick. The formation is named after Moydart Point (2,350 ft southwest of the mouth of McAdam Brook, which is situated on the type section along the coast (Pl. 4).

The formation is divided into two members: the lower green member consists of green mudstones and siltstones with an increasing number of calcareous interbeds toward the top. The upper red member (Fig. 15) consists of marls and concretionary limestones.

The green member has a gradational basal contact with the underlying McAdam Brook Formation. The base is taken where the green mudstone and siltstone beds, many of which are more than a foot thick, become the dominant material (50 ft southwest of shore station 38, Pl. 4).

On the shore, the basal 27 in. of the red member (Fig. 15) consists of thin bands of silty calcareous mudstone, alternately red and green, with thicker sandy bands at the base (Dineley, 1963). It rests with a sharp contact on the siltstones of the green member.

The upper contact of the Moydart Formation is also sharp, although on the shore, the top of the red member (Fig. 15) is irregularly reduced, possibly during a brief period of erosion before the deposition of the overlying Stonehouse Formation.

DISTRIBUTION AND THICKNESS. The Moydart Formation is best exposed on the shore between stations 20 and 38 (Pl. 4), in an outcrop continuous for 1,600 ft. The red member also occurs in McDonald Brook, 300 ft east of its outcrop on the shore. Farther east, only the green member is exposed in McAdam Brook and on the adjacent slopes; the red member is cut out by faulting (Pl. 1).

From a point 1,500 ft east of the mouth of McAdam Brook (Pl. 1) eastward to Smith Brook, both members dip gently to the south in an area cut by several large faults. The best section is in Smith Brook (Pl. 4), where the red member and the top 150 ft of the green member are

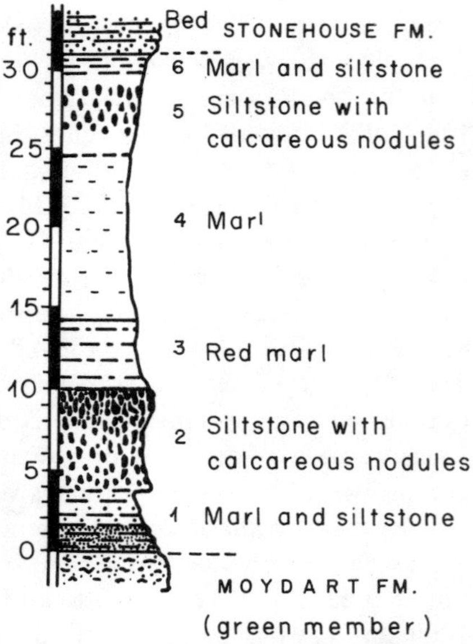

Figure 15. Vertical section of the red Moydart on the Arisaig shore (*after* Dineley, 1963).

exposed north of Route 45. The lower part of the green member is not exposed in this area.

The beds in the lower part of Smith Brook are downthrown to the northwest by a major fault, so that the Moydart beds crop out again in the upper part of Smith Brook (between 2,600 and 3,000 ft south of Route 45). Here the red member (Fig. 15) and the other part of the green member have steep dips (Pl. 1).

In Arisaig Brook (Pl. 7), most of the Moydart beds are exposed, but there are several covered intervals. There are no exposures for more than a mile to the east of Arisaig Brook (Pl. 1). Several small outcrops in the west branch of Doctors Brook (Stations X13 to U5, U15 to U16, U18 to U20, Pl. 5) are assigned to the green member on lithological grounds; no diagnostic fossils have been found in this faulted and folded region near the Hollow Fault.

Apparent thicknesses (in feet) in the best exposed sections are as follows:

	SHORE (PL. 4)	MCADAM BROOK (PL. 8)	SMITH BROOK (PL. 4)	ARISAIG BROOK (PL. 7)
Red member	32	—	20 to 25	15+
Green member	360	150+	150+	650

Of these sections, the shore is the least disturbed tectonically. The faults present have throws of less than 20 ft, and, in most cases, marker beds can be recognized on both sides of each fault. There is some evidence of movement along certain bedding planes, but the estimated thickness is probably within 20 percent of the true figure. Twenhofel's estimate of 347 ft for the thickness of his zones 28 to 31 is in good agreement with the present assessment of 360 ft for the green member of the Moydart on the shore.

This estimate suggests that more than half of the green member is cut out in McAdam Brook, French Brook, and Smith Brook. The large apparent thickness in Arisaig Brook (650 ft) is interpreted as being repetition of parts of the succession by the process of strike faulting.

The only complete exposures of the red member are on the shore (Pl. 4) where 32 ft of beds are visible. The red member (Fig. 15) thins toward the east. In lower Smith Brook (north of Route 45; Pl. 4), 20 ft is exposed, and the maximum possible thickness for the red member (between shelly green Moydart siltstones—collection 10175—and a shelly limestone at the base of the Stonehouse Formation) is only 25 ft.

In upper Smith Brook (Pl. 1; 3,000 ft south of Route 45) less than

20 ft of the red member is visible (Fig. 15). Further thinning toward the east is suggested by the section in Arisaig Brook (Pl. 7, on the valley side north of station 96) where only 15 ft of red beds is exposed; however, the lower contact of the red member was not seen at this point, and the bed is affected by faulting.

The red member is also present in School Brook Cove, Cape George, and in McLellan Brook, Pictou County (Maehl, 1961, p. 65). Moydart brachiopod species have also been found in quartzite in the Cape George fault zone, which contains worm tubes and which indicates coarser sediments in an area to the north of the Arisaig belt.

The green member is also present west of Lochaber Lake.

LITHOLOGY. The green member is composed predominantly of slightly fissile shale or siltstone, with interbeds of massive quartzose limestone and massive calcareous siltstone, and thin shelly limestone. The beds near the base of the green member are composed of 88 percent siltstone, 18 percent massive calcareous beds, and less than 1 percent shelly limestone; and those near the top are: 65 percent shale and siltstone, 25 percent massive calcareous beds, 10 percent shelly limestone. Most of the beds in the green member range between 1 and 12 inches in thickness. A marked increase in the number of thicker beds occurs about 30 to 50 ft from the base of the member (seen on the coastal cliffs between stations 32 and 36, and in McAdam Brook between stations 18 and 20).

The siltstones are greenish gray (5 G 5/1) when fresh and dry, dark greenish gray (5 GY 3/1) when wet, and dark greenish gray (5 G 4/1) when weathered. Locally, subparallel alignment of chlorite and muscovite gives some of the beds a weak bedding plane fissility. A typical siltstone lacks carbonate and consists of about 80 to 85 percent fine silt (0.01 to 0.03 mm) made up of 40 to 50 percent subangular quartz; 10 to 15 percent chlorite; 10 to 15 percent muscovite and sericitized feldspar, and 5 to 10 percent translucent iron oxide; clay-sized matrix makes up the remaining 15 to 20 percent.

The massive calcareous beds have the same range of colors as the shales, except that they weather to light gray (N7). Composition and grain size of these massive beds range between calcareous siltstone and quartzose limestone. The siltstones range in thickness between 2 and 10 in., and are composed of the following: 55 to 65 percent subangular quartz (0.03 to 0.06 mm), 10 to 15 percent muscovite and sericitized feldspar, 10 to 15 percent chlorite, 10 to 15 percent calcite, 5 percent fresh feldspar, 5 percent translucent iron oxide, and 5 percent clay-sized matrix.

The quartzose limestones are usually thinner than the other beds; most are less than 2 in. and all are less than 6 in. thick. Grain size and color are the same as the more quartz-rich beds. A typical limestone has approximately 50 to 55 percent interstitial calcite, 20 to 25 percent subangular to subrounded quartz, 10 to 15 percent interstitial chlorite, 5 to 15 percent sericitized feldspar and sericite, and 5 percent opaque or translucent iron oxide. The calcite, even though it occupies the interstices between quartz and feldspar grains, forms optically continuous units up to 0.3 mm.

All the massive beds have smooth, sharply defined lower contacts. The upper contacts are more variable; some are smooth, some grade upward into shale or mudstone, others show oscillation and interference ripples on the top surface.

The shelly limestones fill scours in the underlying siltstone; they range in thickness from 1 to 6 in. and can be traced horizontally for as much as 50 ft, although most of them extend for less than 15 ft. They are light gray (N7) and contain 10 percent quartz and altered feldspar (0.01 to 0.02 mm), 5 percent muscovite, 5 percent clay-sized particles, and 80 percent calcite (mainly shell material).

These shell beds sometimes grade upward into finely laminated and cross-laminated sandstones, which are well-sorted, pale-brown (5 YR 6/3) beds, usually less than 2 in. but occasionally up to 6 in. thick.

Cross-laminations are particularly well developed in the green member, and wave-eroded shore exposures showing current indicators in three dimensions are abundant. Of fifty measurements, 58 percent showed currents flowing from the northeast quadrant, a similar direction to those in lower formations of the Arisaig Series (Fig. 14).

The upper red member (Fig. 15) on the shore has been described by Dineley (1963). Most of the inland sections show a moderately uniform lithology, mainly red siltstones; but on the shore, a variety of beds are present (Fig. 15).

Beds 1 to 6 contain red micaceous siltstones, ranging in color between moderate red (5 R 4/5), grayish red (10 R 4/2) and dusky red purple (5 RP 3/2). They are mottled with gray (5 GY 7/1) reduction spots, presumably around organic inclusions.

Approximate composition is 70 to 80 percent medium silt (0.01 to 0.04 mm diam), and 20 to 30 percent clay-sized matrix. The silt consists of 30 to 40 percent angular quartz, 10 to 15 percent muscovite and altered feldspar, and 10 to 15 percent chlorite. The iron oxide is interstitial between the clasts and takes the shapes of the voids; it thus appears to be secondary, probably a result of subaerial weathering.

Beds 2 and 5 contain very light gray (N8), ellipsoidal, calcareous nodules in a matrix of red siltstone. Nodules up to 2 inches in diameter by 4 in. long occur in bed 2, becoming more concentrated, and even fused together, near the top of the bed. Bed 5 contains smaller nodules, widely dispersed near the base, slightly more concentrated near the top.

Beds 1, 4, and 6 contain greenish-gray (5 GY 7/1) quartzose limestones and marls.

Approximate composition is 85 to 95 percent medium silt (0.01 to 0.04 mm diam), and 5 to 15 percent clay-sized matrix. The silt consists of 45 to 55 percent calcite, 25 to 35 percent angular quartz, 5 to 10 percent muscovite and altered feldspar, and 5 percent fresh feldspar.

STRUCTURE. The outcrops of the Moydart Formation occur in the least-disturbed parts of the shore section. West of station 34 (Pl. 4), the only signs of movement are along small vertical faults, or along bedding planes in a few places; few fossils are deformed, even orthoceratids longer than 6 in. (common between stations 23 and 25) show no evidence of distortion. East of station 34, however, the beds are more disturbed, there are many small faults and folds, and boudinage is present in some sandstones at station 35. The Moydart beds dip generally south, between 30° and 40°, along the whole of their coastal outcrop and inland to McAdam Brook.

Eastward, the dips decrease. In lower Smith Brook (Pl. 4), only the red member (Fig. 15), just north of Route 45, dips more steeply than 20°.

On the northwest limb of the Arisaig syncline, in upper Smith Brook and Arisaig Brook (Pl. 1), most of the beds are nearly vertical and some are overturned. In Arisaig Brook (Pl. 7), it is probable that many beds are repeated by faulting as the apparent thickness of the beds seen is much greater than that on the shore (*see* discussion on thickness above).

In the west branch of Doctors Brook (Pl. 5), the beds assigned to the green member all young to the east. It is concluded that they lie on the much faulted northwest limb of the Doctors Brook syncline, although exposures are absent to the west of this part of the brook, and it is not possible to be certain about the structural relations with adjacent areas.

PALEONTOLOGY. The green member of the Moydart Formation contains abundant brachiopods, which are especially common in the limestones filling scour depressions. Epifaunal bivalves are also present in these limestones.

The green siltstones contain crinoids [including complete calyces near shore station 25 (Pl. 4), Yale Peabody Museum locality number A3617], infaunal bivalves in living position, and orthoceratids (the latter are especially common between shore stations 23 and 24). Harper (1964) states that, at Arisaig, the following species are confined to the Moydart Formation: *Eatonioides lamellornatus* McLearn, *"Camarotoechia" squamifera* (McLearn), *Sphaerirhynchia saffordi* (Hall), *Rhynchospirina acadiae* Hall, and *Hyattidina northumberlandensis* (McLearn).

The red member on the shore (Fig. 15) contains a few thin phosphatic shells near the base, and some acanthodian spine fragments (?*Onchus*) 14 ft above the base (Dineley, 1963).

The Cape George fault zone near School Brook Cove and Cormorant Cliff Cove, 14 mi northeast of Arisaig, contains some blocks of the Moydart Formation in which Ørvig (1956, written commun., *quoted in* Boucot and others, 1959), records *Thelodus parvidens, Nostolepis* sp., and *"Gomphodus"* sp. of upper Ludlow age ("stratigraphically somewhat below the Ludlow Bone Bed").

Age and Correlation

The underlying McAdam Brook Formation contains early Ludlow graptolites. The green member contains *Hyattidina,* which elsewhere is only known from Silurian beds, and the red member at Cape George (Fig. 15) contains upper Ludlow fish; the age of the Moydart is therefore Ludlow.

The Moydart Formation has also been recognized on lithological and paleontological evidence in Pictou County (Maehl, 1961) and in the Cobequid Mountains (Harper, 1964).

Stonehouse Formation. This formation was named by Ami (1900) but not defined until 1909, when Twenhofel (1909, p. 156) described it: "This subdivision has its base on the top of the 'Red Stratum' and extends to the amygdaloidal trap at the top of the Silurian section." Williams (1914, p. 71) followed Twenhofel in thinking that the shore section "is probably all of the formation that is preserved"; this section corresponds to Twenhofel's zones 33 to 40, and to Williams' zones 11 and 12 (Table 1).

McLearn (1924) recognized that beds younger than any seen on the shore occur along MacEachern Brook (he called it Stonehouse Brook; Table 2); he designated the shore section zones a to c, and the MacEachern Brook section zone d. Our present work is in agreement with Copeland (1964) in suggesting that younger beds of the Stonehouse (equivalent to those seen along McAdam Brook, Pl. 8, stations U8 to U19 and stations U40 and U41) are probably present to the southwest of the youngest bed exposed.

The top contact of the Stonehouse Formation is taken at the top of the highest bed with marine fossils below the nonmarine Knoydart Formation.

This contact is exposed only in one place: station 41 in McAdam Brook (Pl. 8) 3,200 ft southeast of Nova Scotia Route 45. Here there is a conformable gradation from a limestone (with crinoids, ostracods, and bivalves–collection 10207) up through green micaceous sandstones into a nodular limestone similar to those in the red member of the Moydart and to others in the Knoydart Formation.

The base of the Stonehouse Formation is exposed on the shore (Pl. 4, 75 ft east of station 20), in upper (Pl. 1) and in lower (Pl. 4) Smith Brook, and at station 96 in Arisaig Brook (Pl. 7). At all four localities, there is a sharp contact between the underlying nonmarine red member of the Moydart and the basal marine beds of the Stonehouse Formation.

DISTRIBUTION AND THICKNESS. The Stonehouse Formation is exposed on the coast (Pl. 4) between stations 20 and 0; the beds here are faulted in several places, but Twenhofel's estimate of 1,075 ft for the thickness of this shore section is probably accurate to within about 100 ft. At station 0, a Mississippian basalt rests with angular unconformity on red-stained beds which Twenhofel and Williams regarded as the top of the formation. Inland, however, the basal Mississippian rocks rest on younger strata; first the topmost Stonehouse beds, then progressively younger beds of the Knoydart Formation. It is estimated that at least 300 ft of strata are present between the youngest Stonehouse on the shore (Pl. 4, Station 0) and the lowest Knoydart seen on Route 45 (1,750 ft north of McAras Brook; Pl. 1). Of this 300 ft, 200 ft is represented by the Stonehouse exposed in MacEachern Brook (McLearn's zone d), and 100 ft by younger beds not exposed in this brook. A large part, perhaps about 75 ft, of this hidden 100 ft may be equivalent to the youngest Stonehouse beds seen in McAdam Brook (Pl. 8, stations U8 to U19, and stations U40 to U41), where a very reduced brachiopod fauna is present. It is probable that the 75 ft of these beds with a reduced fauna are younger than any of the beds seen along MacEachern Brook which have a normally rich Stonehouse fauna.

An approximate estimate of the thickness of the Stonehouse Formation in the type section along and near the coast (Pl. 4) can therefore be made up as follows:

Shore (stations 20 to 0)	1,075 ft
MacEachern Brook	200
Unexposed beds south of MacEachern Brook	75
Approximate total	1,350 ft

An unknown thickness may be added or subtracted to this estimate due to movement along the thrust fault at the mouths of Stonehouse and MacEachern Brooks; errors of as much as 20 percent may also be present due to structural deformation. Many isolated exposures of the Stonehouse Formation are present between Route 45 and the shore to the west of McAdam Brook, but no continuous inland section is exposed.

Copeland (1964), using ostracodes as zonal indicators, estimated that the basal 600 ft of the Stonehouse is missing in McAdam Brook; this is explained by a fault 100 ft southeast of station L17 (Pl. 8), where the Stonehouse beds are brought into contact with the green member of the Moydart Formation. Upstream from this fault, 700 ft of Stonehouse beds are present on the north limb of a synclinal which has an axial trace between stations U8 and U14 (Pl. 8). The youngest beds seen in this syncline have a reduced brachiopod fauna (similar to the highest Stonehouse between stations U40 and U41) and may be near the top of the Stonehouse Formation. The McAdam Brook section thus suggests a total Stonehouse thickness of a little more than 1,300 ft.

East of McAdam Brook, the basal Stonehouse beds are exposed near Route 45 as far as lower Smith Brook (Pl. 1). East of this point, large faults downthrowing to the west shift the Stonehouse outcrop 3,000 ft to the south as shown by the exposures in upper Smith Brook (Pl. 1).

In Arisaig Brook (Pl. 7), 750 ft of beds are present between stations 96 and 109, and, if no faulting is present, another 750 ft may be present on the ridge 1,000 to 4,000 ft east-northeast of station 96. This 1,500 ft would represent the thickest section of Stonehouse at Arisaig; moreover, the youngest beds seen on the ridge have a full representation of brachiopod species present, and may thus not be quite at the top of the formation. It is most probable that some of these beds have been repeated by faulting.

No exposures of the Stonehouse Formation are present to the east of this ridge.

As there is no section without some covered intervals where the effects of faulting are uncertain, it can only be concluded that the Stonehouse Formation is about 1,350 ft thick, but it might be as thin as 1,250 ft or as thick as 1,600 ft.

Fletcher's (1893) map shows an area of Silurian beds between 5 and 6 mi southwest of Arisaig in the area northeast of Bailey's Brook. Stonehouse fossils (*Isorthis fornicatimcurvata* and *Delthyris rugaecosta*) have been collected from these beds, and the Stonehouse Formation is probably the only part of the Arisaig Group present in the area. The Stonehouse Formation is also present at Cape George, in the Lochaber area, in Pictou County, and in the Cobequid Mountains.

LITHOLOGY. The Stonehouse Formation is similar to the green member of the Moydart, although many beds contain more quartz and less chlorite.

The predominant lithologies are greenish-gray (5 G 5/1) shales, mudstones, and siltstones, ranging in thickness from 1 to 12 in. The compositions of these beds vary considerably as to the amounts of quartz, mica, carbonate, and clay.

The lower half of the Stonehouse consists of about 70 percent slightly fissile shale and mudstone, 25 percent massive siltstone, and 5 percent thin fossiliferous siltstone with thick lenses of shelly limestone. In the upper half of the formation, mudstone and shale make up only 30 percent of the strata and the massive siltstone, 65 percent; the thick shelly limestones are absent, but thin cross-laminated lenses and thin fossiliferous siltstones persist.

The quartz-rich siltstones are usually massive with strong joints normal to the bedding. A typical bed consists of 75 to 80 percent angular quartz, 5 percent feldspar, 5 percent muscovite and altered feldspar, 10 to 15 percent chlorite, and 5 to 10 percent calcite. No clay matrix is present.

Some of these quartzose beds are cross-laminated channel deposits. Oscillation and interference ripples are common in the siltstones. At shore station 5 (Pl. 4), there is a well-exposed channel 2 ft wide and 6 in. deep. Silty mudstones and slightly fissile shales occur in beds as much as 4 in. thick. They are made up of 85 to 90 percent silt-sized (0.007 to 0.05 mm diam) material and 10 to 15 percent clay-sized matrix. The silt is variable in composition, but averages about 50 to 60 percent subrounded quartz, 15 to 20 percent muscovite and altered feldspar, and 15 to 20 percent chlorite. There is a trace of translucent orange iron oxide. No carbonate is present.

Many of the mudstones are massive and contain mica flakes with a random orientation. The presence of burrows and infaunal pelecypods in growth position suggest that the lack of fissility probably resulted from stirring of the sediment by infaunal populations. The surfaces of many mudstones are marked by trails which appear as casts on the under surfaces of overlying siltstones; some of the trails near shore station 9 (Pl. 4) may be trilobite trails (Yale Peabody Museum locality number A3679). Between shore stations 13 and 15, phosphatic lumps are present similar to those found in the Doctors Brook Formation.

Fossils are common in two lithologies which together make up 5 percent of the Stonehouse Formation; thin (less than 1 in.) siltstones and thick (as much as 12 in.) limestones. The thin siltstones can sometimes be traced laterally for more than 100 ft. Their approximate composition is

90 to 95 percent medium and coarse silt (0.02 to 0.06 mm), 70 to 80 percent subangular quartz, 5 to 10 percent calcite-filled shell molds, 5 percent muscovite and altered feldspar, 10 percent chlorite, 5 percent iron oxide and marcasite, and 5 to 10 percent clay-sized matrix. Brachiopods are the most numerous fossils in these beds, but gastropods, bivalves, crinoids, and bryozoans are also common; many of the bivalve shells are unbroken and some are articulated; sections of crinoid stems up to 2 or 3 in. long are present. These beds were apparently deposited by gentle bottom currents, which winnowed out the clay and gently reworked the shells without breaking them or transporting them far.

The thick limestone lenses are 4 to 15 ft across. Their average composition is about 95 percent fine silt (0.01 to 0.02 mm), 75 to 85 percent calcite; 10 to 15 percent quartz, 5 to 10 percent muscovite and altered feldspar, and 5 percent clay-sized matrix.

The limestones occur in scours in the underlying silt. Some grade upward, through cross-laminated very fine sandstones, into siltstones. The shells, 90 percent of which are brachiopods, are more fragmented here than in the thin siltstones; most of them are oriented with their convex side up. The evidence suggests that the sequence of deposition frequently was: shell fragments, fine sand, and then silt. This sequence appears to represent a single cycle of deposition during which the current velocity decreased.

Fifty measurements of cross-laminated units in the lowest 600 ft of the Stonehouse Formation indicate that 54 percent of the depositing currents flowed from the northeast quadrant; this is the same current direction as indicated by the underlying formations. However, 50 measurements on the uppermost 400 ft of beds on the shore indicate that in 60 percent of the cases, depositing currents flowed from the southeast quadrant (Fig. 14). The change in direction is not abrupt; it occurs between 500 and 700 ft above the base of the formation, and coincides with the lithological change from dominant mudstones below, to dominant siltstones above.

STRUCTURE. The Stonehouse Formation on the shore has a consistent southwesterly dip of 35° to 45°, but the sequence contains numerous small dislocations; some of these are normal faults, others are due to bedding plane movements, and some are thrusts on which younger beds override from the south. One of these thrusts, apparently with more displacement than the others, produces a disturbed zone, up to 50 ft wide, which is exposed northeastward from near the Mississippian unconformity on the shore as far as the mouth of Stonehouse Brook (Pl. 1). Northeast from this point, it is conjectured that the outcrop of this thrust plane may be continuous with the thrust in McAdam Brook which cuts out the red member of the Moydart and the basal 600 ft of the Stonehouse near station L17 (Pl. 8).

Upstream from Route 45, the Stonehouse beds are exposed in a syncline with a nearly horizontal axis and gently dipping limbs (stations U8 to U14). Lack of exposures farther upstream hides the relation between these gently dipping beds and the steep southeasterly dips of the highest Stonehouse at station U40.

In upper Smith Brook and Arisaig Brook, the Stonehouse beds dip steeply south or are slightly overturned (Pl. 1). East of Arisaig Brook, the strike swings to a north-south direction, suggesting further structural complications in the axial region of the Arisaig syncline.

PALEONTOLOGY. Brachiopods are the most abundant fossils, although pelecypods, ostracods, crinoids, and bryozoa are all common (*see* Appendix). Some of the mudstones show signs of being disturbed by burrows and contain infaunal bivalves in growth position. Epifaunal bivalves, brachiopods, and ostracods occur in thin siltstones and in limestone lenses.

Copeland (1964) has divided the Stonehouse Formation into three ostracod zones on the basis of the stratigraphic distribution of 15 species and subspecies assigned to 8 genera. On the shore and in MacEachern Brook, his zones are, respectively, 100 ft, 550 ft, and 600 ft thick; this leaves about 100 ft at the top of the formation unassigned (*see* discussion on thickness above).

At Arisaig, the following brachiopods are confined to the Stonehouse Formation: *Isorthis fornicatimcurvata* (Fuchs), *Schizophoria* sp., *Salopina missendenensis* (Straw), *Shaleria gilpini* (Dawson), *Mclearnites mertoni* (McLearn), *Leptaena* aff. *rhomboidalis* (Wahlenberg), *Protochonetes stonehousensis* Harper MS., *Rhynchospirina sinuata* (Hall), *Atrypa* cf. *gedinniana* (Fuchs), *Delthyris (Quadrifarius) rugaecosta* (Hall). Harper (1964) also lists *Proschizophoria* sp. from collections made in the top 50 ft of the Stonehouse Formation in Pictou County (Maehl, 1961).

Most of the brachiopods are distributed throughout the formation, but there is a marked decrease in the number of species present near the top of the Stonehouse Formation at Arisaig. These highest beds are exposed only along McAdam Brook (Pl. 8). At a location 50 ft north of station U20, the usual range of species is present in beds dipping at 30° to the north, but to the north, between stations U17 and U15, three shelly lenses each contain only a limited fauna, mainly chonetids, rhynchonelloids, ostracods, crinoids, and bivalves. Upstream, a further reduction in the fauna is apparent in the limestone immediately underlying the Knoydart at station U41. Here (collection 10207), only 25 ft below a nodular limestone with characteristic subaerial features, the marine fauna is limited to crinoids, bivalves, and ostracods. At the time of deposition of this bed, conditions must still have been marine (crinoids are present), but some environmental factor, perhaps salinity or depth, was excluding all the brachiopods, even those more tolerant forms such as chonetids and rhynchonelloids (which are still present at stations U15 to U17).

The highest beds exposed on the ridge east of Arisaig Brook (Pl. 1) have a diverse brachiopod fauna; therefore, unless the reduced faunas are very local, these beds on the ridge are older than those in McAdam Brook at stations U15 to U17 (Pl. 8). The large apparent thickness (1,500 ft) of the Stonehouse in the Arisaig Brook area is thus almost certainly due to strike faults repeating beds.

The brachiopod diversity does not appear to decrease at the top of the Stonehouse in Pictou County (Maehl, 1961). Dineley (1962) has discovered a single large acanthodian spine in the Stonehouse Formation on the coast section. Conodonts from the same section have been studied by Legault (1966).

AGE AND CORRELATION. The Stonehouse Formation has been determined to be of Pridoli and Gedinnian age by comparison with European sections (Boucot, 1960, p. 291). The presence of *Delthyris (Quadrifarius) rugaecosta* (Hall) throughout the Stonehouse is particularly significant, as *Delthyris* possessing a grooved fold is known only from Pridoli and Gedinnian strata elsewhere in the world. By contrast, only ungrooved forms are present in beds equivalent to or older than the British Ludlow. The presence of *Schizophoria* and *Proschizophoria* is also significant, as neither of these genera is known elsewhere in the world from beds of Silurian age.

Copeland (1964, p. 7) correlates the Stonehouse ostracodes with faunas from Sweden and Germany, which are correlated in turn with the Downtonian of Britain. Legault (1966) reports *Spathognathodus eosteinhornensis* zone conodonts from the Stonehouse Formation coast section. This zone has a latest Ludlow through Pridoli age span.

Dineley (in the following chapter) considers that the fish in the basal Knoydart are of middle or late Downtonian age. Because the fish in question belong to the zone of *Traquairaspis,* it is clear that the underlying Stonehouse must be no younger than upper Downtonian (*see* Fig. 29 for correlation of Downtonian and marine Devonian). In Podolia (Boucot, 1960), it has been previously noted that the zone of *Traquairaspis* overlies a shelly fauna correlated with the lower Gedinnian.

It is concluded that the Stonehouse is of Pridoli and lower Gedinnian age, that is, of post-Ludlow age; this correlates with the lower half of the Downtonian up to a position possibly as high as the zone of *Traquairaspis.*

It should be noted that elsewhere in the world (Boucot, 1960), lower Gedinnian shelly faunas have been found as high as the middle Dittonian, that is, the lower Gedinnian shelly fauna has a proved stratigraphic range of at least upper Downtonian to middle Dittonian, although in Nova Scotia it is restricted to the Downtonian only (Fig. 29).

Arisaig Group

Knoydart Formation. Most of the Paleozoic faunas and some of the formations of Nova Scotia show strong affinities with faunas and rocks of the same age in western Europe. H. Ami (1901b) noted this affinity when he described the Knoydart Formation as a "bit of the Old Red Sandstone of Europe." The present work offers a new description of the formation and emphasizes the similarity that exists between the red-bed Knoydart and the Lower Old Red Sandstone of England and Wales.

Ami (1900a, p. 301) coined the term Knoydart for the red beds of McAras Brook on the Northumberland Strait coastline of Antigonish County, Nova Scotia. The rocks also crop out in Knoydart Brook (Pl. 2) and other streams nearby, and have more recently been found at Lochaber and elsewhere in northern Nova Scotia. Honeyman in 1866 suggested, on general geological grounds, that the barren red rocks above the Stonehouse beds of the Arisaig area were Lower Devonian in age; paleontological evidence was then not available. Twenty years later, however, collectors working for the Geological Survey of Canada obtained vertebrate fossils from the McAras Brook section. Ami sent the fossils to A. S. Woodward at the British Museum, and they were identified as *Cephalaspis* sp., *Onchus murchisoni* Agassiz, *Psammosteus* cf. *anglicus* Traquair, and *Pteraspis* cf. *crouchi* Lankester. These species were known in the Lower Devonian Old Red Sandstone formations of the Welsh Borderland, and Woodward had no hesitation in suggesting a correlation between the Knoydart Formation and the "Lower Old Red Sandstone-Cornstones." E. I. White (1935) showed that the pteraspid fossils were not *P. crouchi,* but a distinct smaller form. The species is closely related to *P. (Simopteraspis) leathensis* White, which occurs stratigraphically below *P. crouchi* in Britain. The *Psammoteus* cf. *anglicus* listed above was almost certainly a misidentification of a fragment of *Traquairaspis*. The latter genus has now been identified in the Knoydart Formation (Dineley, 1964). The material sent to Woodward appears to be lost, and has not been re-examined. From a bed slightly higher than that from which the original pteraspid came, R. H. Denison obtained a fine collection of a small *Pteraspis,* later described (Denison, 1955) as *P. whitei*. More recently, a locality near Ardmore has yielded abundant material which, although in part distorted by the local cleavage, can be identified unhesitatingly as *Pteraspis (Belgicaspis) crouchi* Lankester (Dineley, 1967). The discovery of this species, a zonal fossil in the British Dittonian, emphasizes the close faunal affinities of the British and Canadian formations.

Denison (1955, 1956) rejected Ami's view that the deposits were those of a lake, and suggested that the beds were laid down off the mouth of a large river—a deltaic region where the dividing line between marine and nonmarine environments is difficult to draw. The present work supports the idea that they were deposited on the flood plain of a delta, probably very near to the sea. Fluviatile agencies are apparent and the sedimentary features of the Knoydart Formation correspond so closely to those of the Downtonian and Dittonian rocks of England and Wales that the over-all conditions must have been the same (Ball and Dineley, 1961; Allen and Tarlo, 1963; Allen, 1963).

Reference to the Knoydart Formation was made by Maehl (1961) when describing the Silurian of Pictou County, and some of the petrography was included.

DISTRIBUTION AND THICKNESS. By far the best section of the Knoydart Formation occurs in McAras Brook (Table 4), but McAdam Brook (Pl. 1) and Knoydart Brook also offer good exposures. Strata are revealed at many isolated points over the outcrops in streams and in road cuts and isolated crags on the higher ground. In Pictou County, some 250 ft of the formation are exposed in the McLellan Brook area, but further exposures are few and poor.

LITHOLOGY. The Knoydart consists essentially of hardened red (5 R 3/2 to 5 YR 4/2) and green (10 G 5/2 to 5 BG 5/2) silty mudstones or marls alternating rapidly with hard sandstones and occasional interclast conglomerate bands. Top and bottom contacts are still somewhat debatable. As far as can be seen in the McAras Brook area, a thickness of at least 1,000 ft must exist. Fletcher's record of the section (Ami, 1901), as he realized, is incomplete, showing some 683 ft only. Since the range of lithologies does not vary from one part of the succession to another, it is not possible to subdivide the Knoydart into smaller lithological units. The few known fossil localities are broadly scattered across the formation and suggest a possible biostratigraphy like that of the upper Downtonian and Dittonian of Britain.

It has often been assumed that the Knoydart Formation rests unconformably upon the Stonehouse Formation of the Arisaig Group. However, as shown in the previous chapter, this underlying series extends strati-

TABLE 4. Section on McAras Brook, Based

	FT	IN.
60. Shale, red, much altered	6	0
59. Trap	4	0
58. Siltstone, green, argillaceous, the upper part metamorphosed	2	0
57. Siltstone, red	18	0
56. Sandstone, green and red, massive	11	0
55. Siltstone, red; fossils	8	0
54. Sandstone with clay pebbles, micaceous, gray and red; fossils	5	0
52-53. Shale, bright red, soft, with some silty and calcareous bands; fossils	30	0
50-51. Shales and siltstone, red, with occasional nodules	30	0
49. Sandstone, red, micaceous in two layers; fossils	4	0
48. Shale, red	3	0
47. Sandstone, siltstone, red green (No. 7)	4	0
46. "Reddish altered rock at the level of the road under the schoolhouse, not well seen"; siltstone	20	0
45. Siltstone, red and green	(?)20	0
44. Conglomeratic siltstone, green, calcareous "from which Dr. Ami obtained the fish remains, *Pteraspis*, etc. first sent to Dr. Woodward (No. 6)"	0	6
43. Siltstone red, hard	10	0
42. Shale, red and green	15	0
41. Siltstone, red, with green layers in "cliffs at road"	30	0
40. Shale, red, green-blotched, nodular in part. Bridge over Route 45 ("shore road") here	31	0
39. Siltstone, red, forming waterfall	14	0
38. Shale, red, silty	15	0
37. Shale, red, fine "The top comes to the foot of the falls in a gorge from which Mr. Weston is supposed to have obtained his fish remains." (No. 5)	12	0
36. Siltstone, gray-green	3	0
35. Sandstone, fine, quartzose, gray, "over which the little brook from the westward falls into main stream at water level"	3	0
34. Siltstones and shale, green, with plant remains	3	0
33. Shale, green, "at mouth of little brook to westward (No. 4); from this the seeds and plants were obtained by Dr. Ami in 1896"	2	6
32. Red siltstones and shales	6	0
31. Shale, fine red with green blotches	5	0

on Fletcher's Notes, in Descending Order

	FT	IN.
30. Sandstone, red, massive	6	0
28-29. Shale, red, siltstone intercalations	42	0
26-27. Shale, red, crumbly but with harder bands	21	0
25. Siltstone and shales, red and green	32	0
24. Shale, red, mottled green	8	0
23. Sandstone, green massive	4	0
22. Sandstone, greenish, hard, siliceous, "containing much carbonaceous matter and cut by veins of quartz" (No. 3)	3	0
21. Shale, red, with siltstone intercalations	14	6
20. Shale, red, crumbly	2	0
19. Siltstone—very fine sandstone, red, micaceous	1	6
18. Shale, red, argillaceous with thin sandy intercalations	5	0
17. Siltstone, red and green, massive, calcareous; locally with interclast shale pellets and nodules "fish remains"	7	0
16. Sandstone, red and green, calcareous, "from which Dr. Ami collected many fossil fish remains in 1897. The upper part contains broken carbonized plants, fish, etc. (No. 2)"	2	0
15. Shale, silty red	5	0
14. Sandstone, fine, red and green "with fish remains"	11	6
13. Siltstone, red, purple, massive	6	0
12. Shale, red, argillaceous, blotchy	22	0
11. Sandstone, green, micaceous silty and parallel bedded with small pyrite nodules. Fossil plants and fish remains (No. 1)	4	0
10. Siltstone, red, shaly, blotched in many places	46	0
9. Sandstone, red parallel bedded	3	2
8. Shale, red, silty	8	0
7. Sandstone, silty, calcareous, red and green flaggy	1	0
6. Siltstone-shale, red	7	6
5. Siltstone, red, highly calcareous and nodular, at mouth of little brook from east	1	6
4. Shale, red	6	0
3. Siltstone, red, shaly, with green-gray blotches and streaks	6	0
2. Siltstone, red argillaceous	4	0
1. Shaly siltstone, red	3	0
Concealed interval 30 ft		
Igneous rock		

graphically up into the lower Gedinnian (or Downtonian) stage of the Lower Devonian. Exposures in McAdam Brook (Pl. 8) reveal a conformable junction between these units (station U41 and slope to west). Greenish calcareous siltstones with brachiopods give way to unmistakable Knoydart lithologies without any sign whatever of a tectonic or stratigraphic break. Mapping and general field relations also suggest that the two units are conformable.

Upper Downtonian fossils occur some 240 ft above the lowest beds of the Knoydart section in McAras Brook and the uppermost Stonehouse fossils have Gedinnian affinities. Any part of the succession that is missing here must be very small. There is, moreover, in the lithological succession a progressive and uniform swing from the calcareous thinly bedded Stonehouse siltstones to the more pelitic Knoydart beds. The first conspicuous sandstone unit of the Knoydart in McAras Brook is some 80 ft from the base (Table 4).

The upper limit of the Knoydart Formation is marked by the unconformity beneath the local Carboniferous rocks.

Allen and Tarlo (1963; Allen, 1963) have used the term "cyclothem" to denote the unit of an intraformational conglomerate followed by sandstone and then siltstone with concretionary bed, found so often in the British Lower Old Red Sandstone. These units vary in thickness from a few feet to tens of feet; below each, there is an erosion surface. To Ball and Dineley (1961), these units seemed to be so variable in lithology and thickness that it was not considered possible to trace them with any certainty over wide areas and use of the term "cyclothem" was avoided. The Knoydart beds are just as variable. From the exposures available, one can determine the presence of many such sandstone-siltstone units, commonly although not always, with the intraformational basal conglomeratic layers and erosion surfaces. Conglomeratic layers average about 2 percent of the succession, the sandstones about 30 percent. This may have originally been even smaller, for the rocks have all suffered a slight degree of deformation which has given a "pencil" cleavage to the siltstones and mudstones and has hardened and fractured the arenaceous beds.

The erosion surfaces below the sandstones are sharp and uneven, usually 1 to 5 inches in relief over as much of the surface as can be seen. Above the erosion surface, scattered or closely packed "marl," mudstone, or siltstone intraclast pebbles may be present together with calcareous and ferruginous hazelnut-sized concretions ("race"). The sandstone commonly shows distinct grading up into finer material with a variety of bedding and other structures. The upper surface of the unit is usually indistinct, grading with the overlying siltstone, mudstone, or "marl." Often the latter is shot through with small calcareous concretions in its upper parts but is otherwise not a true marl.

Fine-grained sediments (argillites or lutites) compose 50 to 75 percent of the strata. They are extremely variable in composition and physical characters. C. F. Hickox has calculated an approximate mean composition of the argillites as 45 to 55 percent silt (0.008 to 0.06 mm diam) and 45 to 55 percent clay-sized matrix. The silt component consists of 15 to 25 percent chlorite, 10 to 15 percent quartz and altered feldspar, 5 to 10 percent muscovite, clastic; 5 to 10 percent opaques (marcasite, organic carbon, and/or iron oxides), and 0 to 15 percent calcite.

The matrix of clay sizes appears to be a quartz-mica mixture rather than predominantly of clay mineral content.

The sandy beds (arenites) are colored either green or red depending upon ferric oxide content. They are extremely variable, but the most common type consists of fairly well-sorted angular to subrounded sand-sized particles. Many are arkosic, and Hickox suggests a general composition of: silt (avg diam 0.08 mm), 50 to 60 percent quartz (subangular), 25 to 35 percent sericitized feldspar, and a trace of plagioclase (fresh, striated). Other constituents are 5 to 15 percent ferric oxide and marcasite, 5 percent chlorite, 5 percent rock fragments (quartz and shale), and 5 percent opaque organic carbon.

Two types of ferric oxide occur in these arkosic sediments: (1) a hematite pellicle on quartz grains and (2) amorphous sand-sized particles of translucent hematite or triogite which is concentrated in the darker-colored lamellae.

Many of the sandy beds show a poor variety of graded bedding with conspicuously coarse basal layers up to a few inches thick and often with shale bleb intraclasts. Indeed, many grade vertically and laterally from intraclast conglomerates. In these conglomerates, the pebbles or pellets are of several different kinds: (1) intraclast or rip-up fragments, flat and rounded or subangular and up to 6 inches in diameter, (2) small calcareous rounded (intraclast) nodules, probably washed from penecontemporaneous Knoydart beds, (3) infrequent highly ferruginous nodules about walnut size or smaller, and (4) rare quartz or other exotic rock pebbles, well rounded and generally less than 1 inch in diameter. The amount of matrix in the conglomerates varies greatly, but the basal layers generally consist of closely packed pebbles. These characteristics are very like, if not identical to, those of the conglomerate beds in the Ditton Series of England and Wales. Even the coarsest of these beds have been found to yield both vegetable carbon and fragmentary ostracoderm remains (as in the crags on the hillside above Ardness).

There is an extraordinary correspondence between the primary and penecontemporaneous depositional structures found in the Knoydart Formation and the Dittonian strata of the Anglo-Welsh area. These features have now been described in sufficient detail by various authors (especially Allen, 1960, 1963). They are well displayed in virtually all good exposures of the Knoydart Formation, especially at McAras Brook and at Lochaber. The primary structures figure most conspicuously in these sections; along McAras Brook, no penecontemporaneous slump or convolute structures have been seen. Some "curly-bedding" is present, however, at Lochaber. No load casts or slump balls have been found so far.

On the other hand, there are some very large and impressive plugged pipes (Allen, 1961) in the section south of Route 45 at McAras Brook. Trace fossils, too, here are locally very common. They consist of both sinuous vertical burrows, round in cross section, about one-half inch in diameter, and up to 5 in. long. Surface tracks and trails suggesting small annelids rather than arthropods, and about one-quarter inch in width, are present on a few bedding planes here, but have been found nowhere else.

Allen (1963) produced a useful table to compare the Dittonian facies in Pembrokeshire and the Welsh Borderland with the modern sediments described by McKee (1938, 1939) and by Sykes (1937a, 1937b) from the Colorado River. The idea is used again below to include the Knoydart Formation (Table 5).

Flow directions for 50 depositing currents have been measured on cross-bedding and ripple marks in the rocks along McAras Brook. Of these currents, 64 percent flowed from the southeast quadrant, as did those in the upper part of the Stonehouse Formation. However, this direction is 90° different from the northeast flow, so prevalent in all lower formations (Fig. 14).

PALEONTOLOGY AND CORRELATION. Fossils thus far discovered in the Knoydart Formation include ostracoderm and placoderm fishes, ostracods, and plants (algal and vascular). On the whole, they are not well preserved because of the induration and slaty cleavage developed in the rocks. No eurypterids have been found during this study.

Fish remains occur in exactly similar beds to those found in the British Dittonian. Of the three most productive beds, one is a thick intraformational conglomerate, two are sandstones with thin basal conglomeratic layers. Scattered and distorted pteraspid remains are also found in the red silty shales of McAras Brook between fossil horizon numbers 50 and 53 (Table 4).

As mentioned above, Smith Woodward identified *Cephalaspis* sp., *Onchus murchisoni* Agassiz, *Psammosteus* cf. *anglicus* Traquair, and *Pteraspis* cf. *crouchi* Lankester. It is not certain that they all came from the one horizon, and the locations of other material from the several previously recorded fossiliferous horizons within the McAras Brook section is not known. R. H. Denison described the fauna from horizon 49 which includes *Pteraspis whitei* Denison, *Cephalaspis,* and *Onchus.*

From a red siltstone located near McAras Brook and probably some feet below Denison's horizon, the author obtained a large fauna which includes large fragments of *Traquairaspis symondsi* (Lankester) (Dineley, 1964), cephalaspid fragments, anaspid scales, *Onchus murchisoni* Agassiz and other vertebrate fragments. No pteraspid occurs with these fossils, and they suggest a high Downtonian level. Denison's pteraspids, occurring close above, appear to correlate well with *Pteraspis (S.) leathensis* White at the base of the Dittonian.

TABLE 5. Sedimetary Features in the Knoydart Formation and the Old Red Sandstone

SEDIMENTARY FEATURE	ANGLO-WELSH O. R. S. FORMATION	KNOYDART FORMATION
1. Intraformational conglomerate	C	R-C
2. Fine sandstone	A	A
3. Very fine sandstone	A	A
4. Siltstone	A	A
5. Concretionary limestone	C	R
6. ?Tuffaceous beds	R	R
7. Ripple-drift bedding (no discordance)	R	R
8. Ripple-drift bedding (with discordance)	A	C
9. Linguloid current ripples	R-A	R
10. Parallel current ripples	R-A	R
11. Oscillation ripple marks	R	R
12. Flat-bedding	A	A
13. Primary current lineation	C	C
14. Planar cross bedding	C	C
15. Trough cross-bedding	C	C
16. Sun-cracks	A	NK
17. Rain prints	R	NK
18. Scoured surfaces	A	A
19. Large channel structures	R	NK
20. Sandstone pipes	A-C	R
21. Slumped bedding	R	NK
22. Animal burrows and trails	A	A

Horizon 53, a bright brick-red spotty siltstone, has yielded pteraspid fragments of a species larger than *P. whitei* and possibly resembling *P. crouchi*.

From horizon 54, about 50 more or less complete dorsal discs of a pteraspid and some ventral discs have been collected. There is some small fragmentary pteraspid material in this faunule, and the median discs have clearly been gathered together by winnowing or some other *post mortem* agency. Cephalaspid head shield rims and cornual fragments number about 15. These are the highest fossiliferous beds found in the McAras Brook section and the species present suggest strongly that the beds be correlated to the *P. crouchi* Zone of the British Dittonian. A similar correlation is suggested by a faunule from the hill-top crags near Ardmore; this locality must be near the top of Knoydart succession.

Thus the faunal evidence now available strongly supports the view that the Knoydart Formation extends from the middle or upper Downtonian to the middle Dittonian.

Structural Geometry and History

INTRODUCTION

Although possessing an apparent over-all structural simplicity, the rocks of the Arisaig area bear evidence of five phases of deformation (D_1...D_5), excluding faulting, and pose a number of vexing structural problems. The area is structurally bounded to the southeast by the Hollow Fault zone (Pl. 1; Fig. 16); this is probably a zone of repeated strike-slip and dip-slip movement. The highland country of Antigonish County, lying southeast of the Hollow Fault zone, is a low-grade metamorphic tract of Ordovician rocks, the Browns Mountain Group. The Hollow Fault zone appears to extend northeastward and southwestward for some miles, and it has been suggested (Wilson, 1963) that it may be part of a fundamental fault zone extending beyond the limits of Nova Scotia. Few minor folds can be seen in the area, but cleavages and lineations are well developed, and a detailed study of these has permitted a tenable structural hypothesis to be constructed.

D_1 STRUCTURES

The dominant and most persistent fold in the Arisaig area is the upward-facing Arisaig Syncline, the axial trace of which passes northeast-southwest across the southeastern part of the area (Fig. 16). Northeast of Arisaig Brook, the northern, southward-facing limb of the fold is generally overturned and the southern, northward-facing limb of the fold is gently inclined to the west, or northwest (Fig. 17). The regional fold plunge is gently to the southwest, but local, shallow, plunge-reversals occur in major parasitic folds (for example, the Doctors Brook Syncline) on the southern limb. A clear structural pattern emerges in the stereographic plot of D_1 structural data (Fig. 18B) in the Doctors Brook area. Bedding pole concentrations, in a strong π girdle, represent the northern and southern limbs of D_1 folds. A penetrative slaty cleavage (S_1) in pelitic formations grades into a less penetrative cleavage in the coarser silty formations, and represents a generally northwestward-dipping axial surface to D_1 folds, and produces a ubiquitous intersection lineation (L1) with bedding (Fig. 18A). Two D_1 parasitic pod-fold pairs, the Ross Brook and McAdam Brook

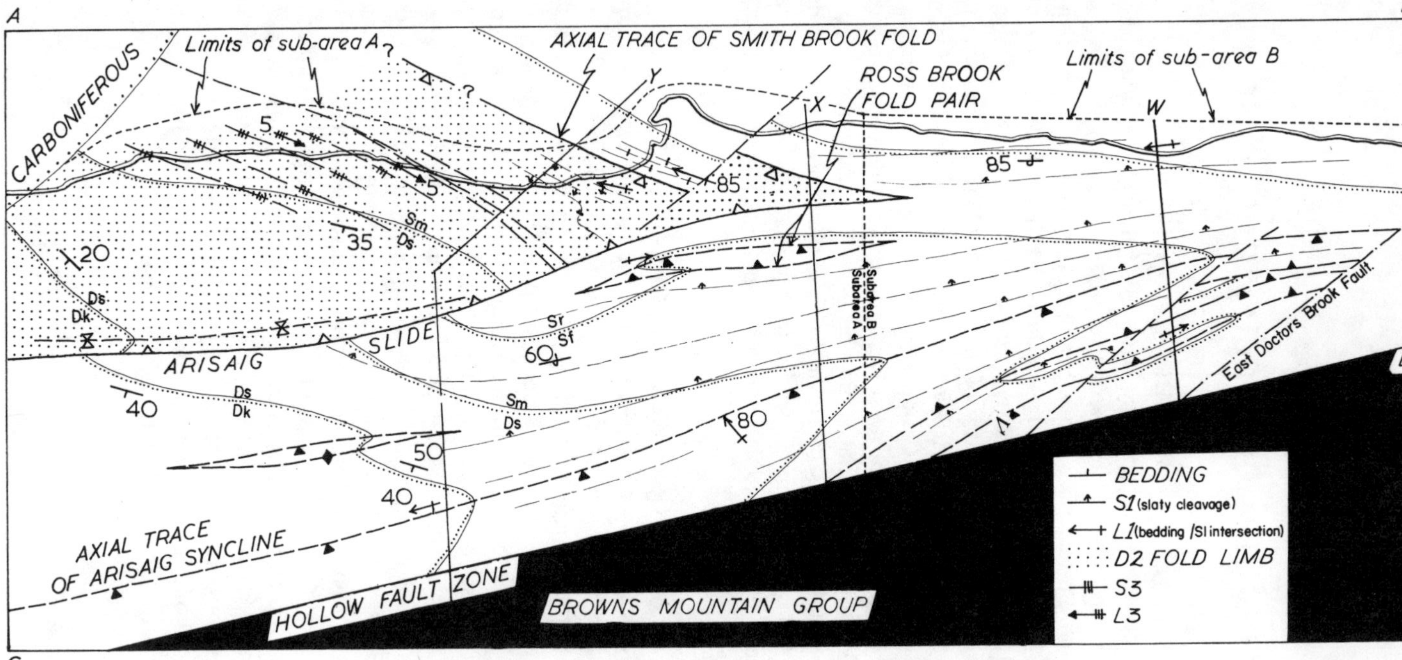

Figure 16. Outline structural map of the Arisaig area.

Figure 17. Structural sections across the Arisaig area. Lines of section, W,X,Y, shown in Figure 16. Arrows indicate facing on S planes. See Figure 16 for legend.

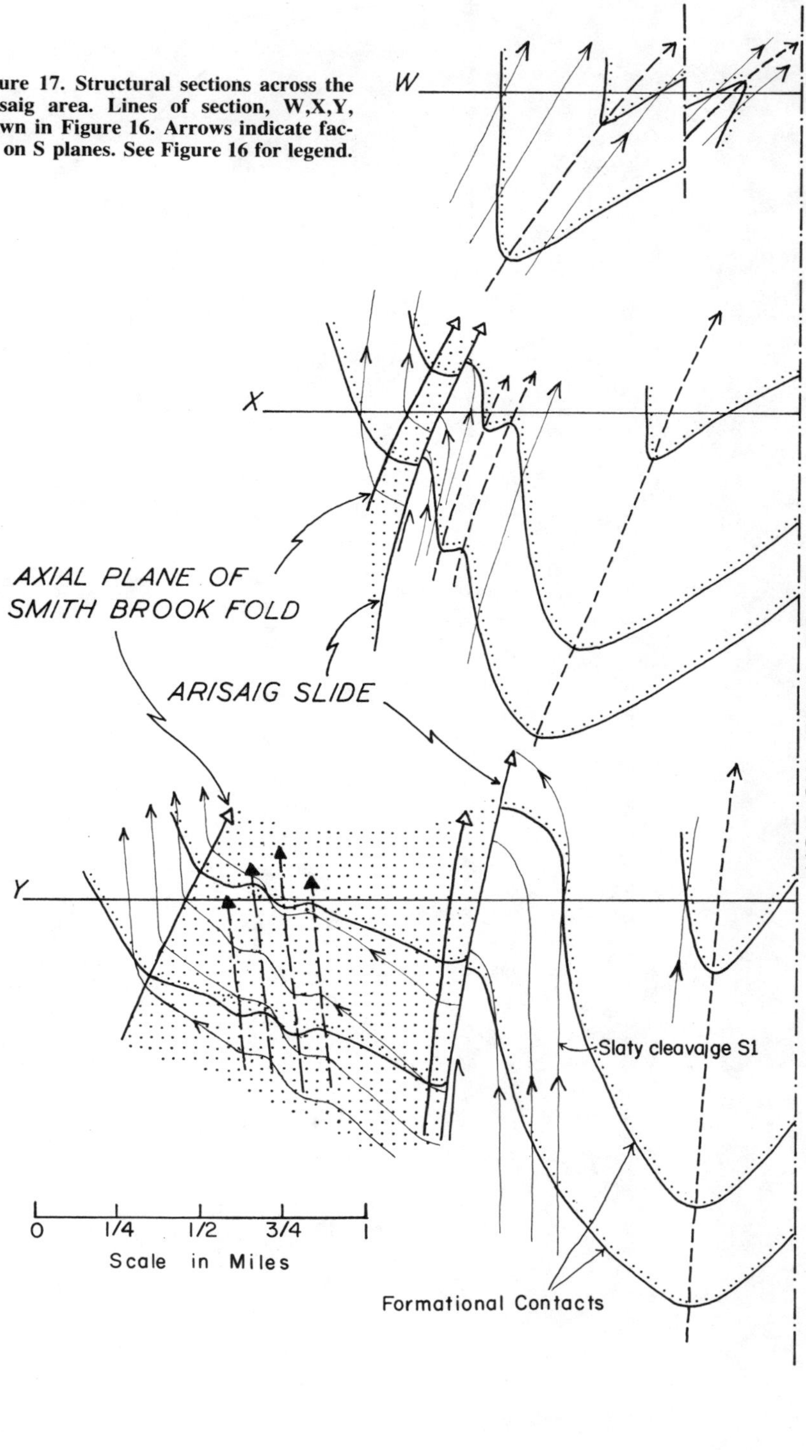

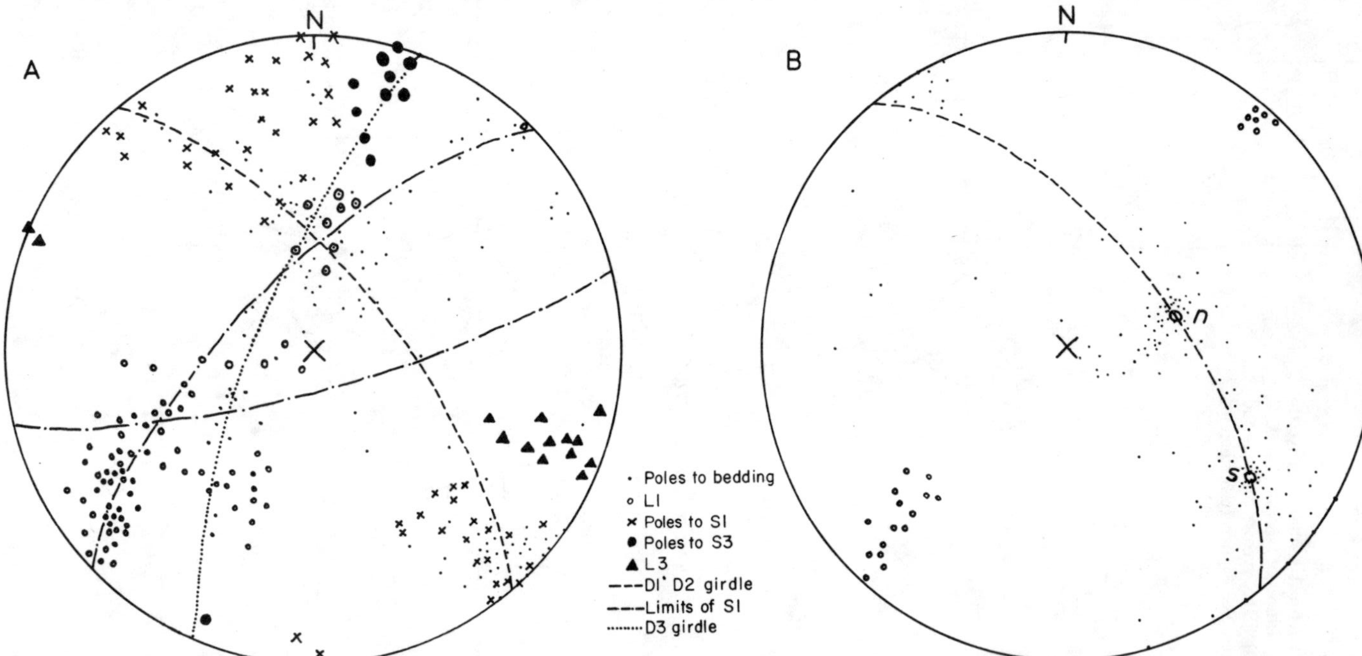

Figure 18. Stereographic projections of pre-D4 structural data in (a) subarea A and (b) subarea B. Limits of subarea shown in Figure 17.

fold pairs, affect the northern limb of the Arisaig Syncline and the northern limb of the syncline swings to an east-west strike in the southwestern part of the area. Concurrent with this strike-swing, there is a gradual opening of the fold, a slight increase in plunge, and an increasing degree of symmetry across a near-vertical axial plane. On the plateau west of Arisaig Brook, the pitch of the Arisaig Syncline is 90°, imparting here a neutral westward-facing attitude to the syncline. A similar local steepening of D_1 meso-folds into neutral westward-facing attitudes is seen on the coast in the Ross Brook Formation between Ross Brook and Arisaig Brook.

No D_1 minor folds have been found in the Arisaig area, few hinges of D_1 folds are exposed (one, for example, is seen in Doctors Brook, Pl. 5, station M8-9), and there is too little detailed control on formation boundaries to permit profile analysis of the D_1 major structure. It is certain, however, that the style is not concentric and the map when viewed southwest down the average plunge reveals the general shape of the D_1 major folds. Indeed some flattening, evidenced by the slaty cleavage and distorted oöliths in the hematite band of the French River Formation, in Arisaig Brook, must have occurred during D_1, and the over-all fold style possesses concentric and similar elements. The long axes of the deformed oöliths pitch northeast in S_1 at between 60° and 80°, and it is likely that the local large variations in the attitude of L_1 are due to variable stretching in S_1, in the manner described by Dewey and McManus (1964). Variable stretching was also the probable causal factor in the development of a slide on the axial surface of the Doctors Brook Anticline.

D_2 STRUCTURES

The major structural geometry in the area, so far described, is regarded as a D_1 geology essentially unmodified, except on a much smaller scale, by later movements and is left unornamented in Figure 16. The stippled area on Figure 16, however, is characterized by bedding dips of between 20° and 45° to the south. This area, well exposed on the coast southwest of Arisaig Brook, is bounded to the south by a major northeast-trending slide, the Arisaig Slide (Figs. 16 and 17). The Arisaig Slide dies out southeast of Arisaig village, downthrows to the north, and has a maximum throw in the region of Smith Brook. The area of gentle southerly dips is bounded to the north by the axial trace of the Smith Brook Fold (Figs. 16 and 17). Within the stippled wedge-shaped area, S_1 dips to the south at a slightly greater angle than does the bedding (Fig. 17), except where it is refracted in the coarser lithologies. Thus the Arisaig anticline and Slide and Smith Brook Fold clearly fold bedding and S_1, and the region between the axial traces of these folds is regarded as having been rotated, from a steep attitude on the northern limb of the D_1 Arisaig Syncline, back into a gently inclined attitude. The northerly downthrow on the Arisaig Slide is consistent with a northward sense of rotation during D_2.

D_3 STRUCTURES

The results of the third phase of deformation are of limited recognized extent. A roughly vertical, nonpenetrative cleavage (S_3), striking east-west, produces a horizontal or gently eastward-dipping intersection lineation (L_3) with bedding, on the coast section between Arisaig Brook and Stonehouse Brook. Associated open D_3 folds are rarely developed. Stereographic projection of structural data (Fig. 18) in subarea A emphasizes a composite D_1-D_2 bedding pole girdle and a weak D_3 girdle.

D_4 STRUCTURES

Structures developed during D_4 are of wide extent and consist of box folds, conjugate folds, kink bands, mitre folds, and intrafolial generative folds. Unlike the D_1 folds, the D_4 folds and associated structures do not affect all outcrops, and thus the effects of D_4 are not penetrative. D_4 mesofolds are most common in Doctors Brook in the pelites of the Ross Brook, French River, and Doctors Brook Formations; they also occur on the coast and inland west of Arisaig but the hinges are rarely well exposed. D_4 folds are best developed on the northern steeply dipping limbs of D_1 folds and consist in this situation, of steeply plunging Z- and S-shaped folds. Occasionally the Z and S folds are in symmetrical, orthorhombic, conjugate arrangement (Fig. 19A), but (more commonly) interstratal, generative Z folds predominate over open S-shaped kink bands (Fig. 19B) or larger flexures. Generally, a direct conjugate relation between Z and S folds cannot be established, but the fact that they are statistically coaxial suggests that they are also coeval and that the intermediate strain axis lay in foliation (Ramsay, 1962). Where exposures permit observation, it is evident that the (dominant Z) folds affect relatively few layers. They are interstratal and generative (Doctors Brook, Pl. 5, stations M2, M26, L31, U65). The largest Z fold exposed (Doctors Brook, Pl. 5, station MO) has a short limb more than 150 ft in length, but limbs of 10 ft or less are common, and most folds affect less than 100 ft of strata. The Z folds are thus asymmetric, the short limbs striking southeast and their vertical or steeply eastward-dipping axial-planes striking just east of north across the regional D_1 strike. They appear to have been developed by a dextral rotation (averaging 90°) of the short limbs. The intrafolial nature of the folds

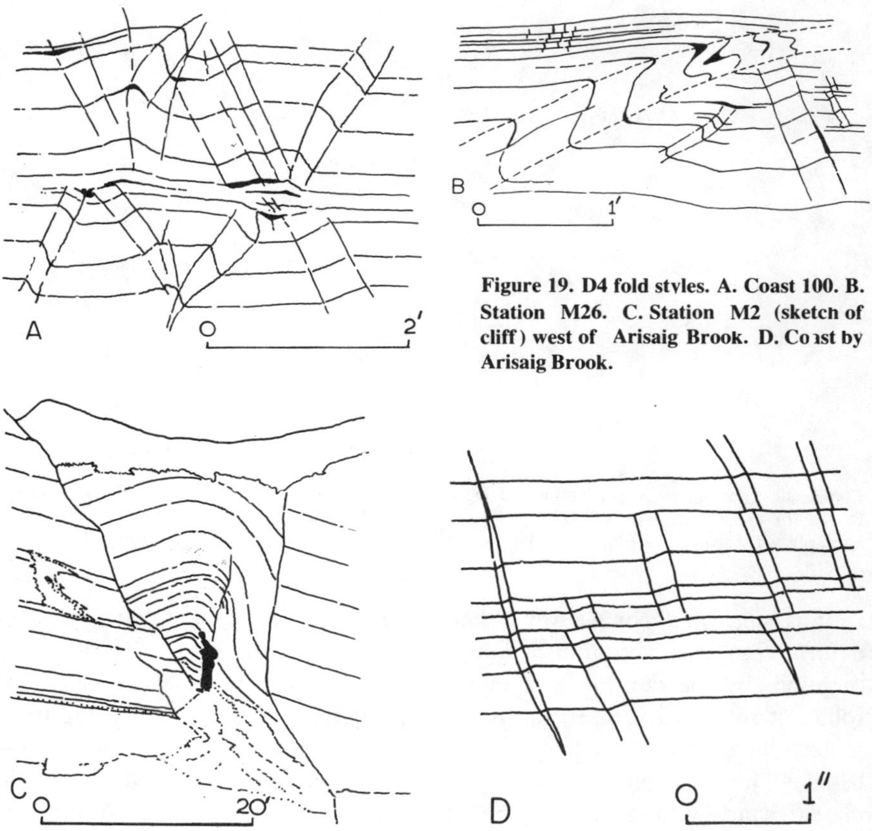

Figure 19. D4 fold styles. A. Coast 100. B. Station M26. C. Station M2 (sketch of cliff) west of Arisaig Brook. D. Coast by Arisaig Brook.

suggests that the dominant dextral rotation may have been accompanied by dextral slip on foliation. In one locality (Doctors Brook, Pl. 5, station M2), quartz veins crossing bedding, unrotated by nearby Z folds, are displaced along bedding planes in the suggested dextral sense.

A few D_4 folds are S-shaped on steeply inclined D_1 limbs. Some occur adjacent to Z folds to form conjugate box folds but more frequently occur as sets of kink-bands (Fig. 19D) or as open flexures. In spite of local orthorhombic symmetry in the Z-S conjugate system, monoclinic symmetry, with dominant Z folds, is far more common. On the basis of the change in form, and crosscutting relations elsewhere in Nova Scotia, one of the authors (Fyson) believes that the kink bands are not conjugate to the Z folds, but formed later. Stereographic projections of D_4 structural data are given in Figure 20.

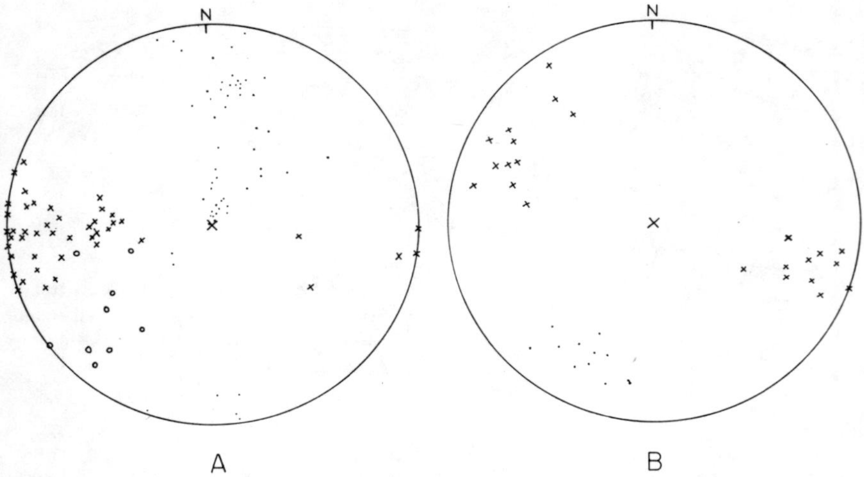

Figure 20. Stereographic projections of D4 structural data. A. In the Doctors Brook region. B. On the coast southwest of Arisaig. Crosses = axial planes of dextral folds; open circles = axial planes of sinistral folds; dots = L4.

It is suggested (by Dewey) that, on steeply inclined D_1 limbs, the D_4 deformation was accomplished, in response to northeast-southwest compression, by the development of conjugate fold systems. Dextral shear in foliation rotated σ_1 and σ_3 (principal and least-stress axes) in an approximately horizontal σ_1-σ_3 plane, and thus monoclinic symmetry was established. This relation has been demonstrated by Hubbert (1951) in his classic sand-faulting experiments and in naturally occurring kink-band systems by Dewey (1965). Another explanation (suggested by Fyson) is that the dominant Z folds with steep north-south axial planes were formed by east-west compression, and that the kink bands formed later in planes oblique to the compression (compare with Fyson, 1966).

D_4 folds superimposed on the southern gently dipping limbs of D_1 folds in Doctors Brook are rare and consist of either open-to-closed (Fleuty, 1964) meso-folds whose axial planes strike just east of north and dip at between 40° and 50° to the east or minor symmetrical conjugate fold systems.

D_4 folds on the coast west of Arisaig, on the southern limb of the D_2 Smith Brook Fold, plunge to the south in axial planes dipping at between 40° and 50° to the east. Small east-dipping thrusts that pass into bedding planes are associated with some of these folds and suggest that D_4 folding was accompanied here in part by the movement of upper beds westward. Just west of the mouth of Arisaig Brook a large D_4 concentric buckle fold is contained in mitre-fashion (Hills, 1963) between divergent bounding planes (Fig. 19C) and plunges south parallel to the dip of bedding, indicating an east-west shortening.

The over-all kinematic pattern indicated by the results of D_4 is (according to Dewey) northeast-southwest compression along the regional D_1 strike, causing the development of conjugate fold systems. This compression was accompanied by dextral shear evidenced by the monoclinic symmetry of conjugate systems superimposed on steeply inclined D_1 fold limbs. Possibly there was dextral movement along the Hollow Fault zone during D_4. Bulk extension on steep D_1 limbs was horizontal and on gently inclined D_1 and D_2 limbs essentially vertical. Quasi-flexural (Donath, 1964) and flexural mechanisms of deformation predominated during D_4 and kinematics were locally controlled by the orientation of foliation, generally bedding in the siltstones and S_1 in the pelites. Thus bulk extension was horizontal where foliation was steep and vertical where foliation was gently inclined.

D_5 STRUCTURES

D_5 structures are rarely seen as kink-bands with horizontal or subhorizontal axial planes. Several D_5 kink-bands in Doctors Brook occur in conjugate pairs (Doctors Brook, Pl. 5, stations L3, M11); the diverging axial planes, when projected, intersect in a subhorizontal line and the kink-band shape indicates shortening approximately down the dip of bedding. Similar kink-bands deform subvertical bedding in Arisaig Brook, some passing into lag faults. In general, D_5 can be related to bulk vertical shortening and northwest-southeast extension.

SUMMARY OF STRAIN SEQUENCE

The major structure of the Silurian/Devonian rocks in the Arisaig area was produced during D_1, except for the gently inclined D_2 fold limb between the Arisaig Slide and the axial trace of the Smith Brook Fold. Figure 21 summarizes the geometric relations of structure and bulk strain developed during successive phases of deformation.

The D_1 phase resulted in a ubiquitous penetrative deformation. Flattening and cleavage development were preceded by major stratal buckling, and the striking local variation in the attitudes of D_1 folds was probably the result of differential stretching in S_1. Bulk strain was a northwest-southeast shortening with near-vertical extension.

The D_2 phase produced the northward rotation of a wedge-shaped flap

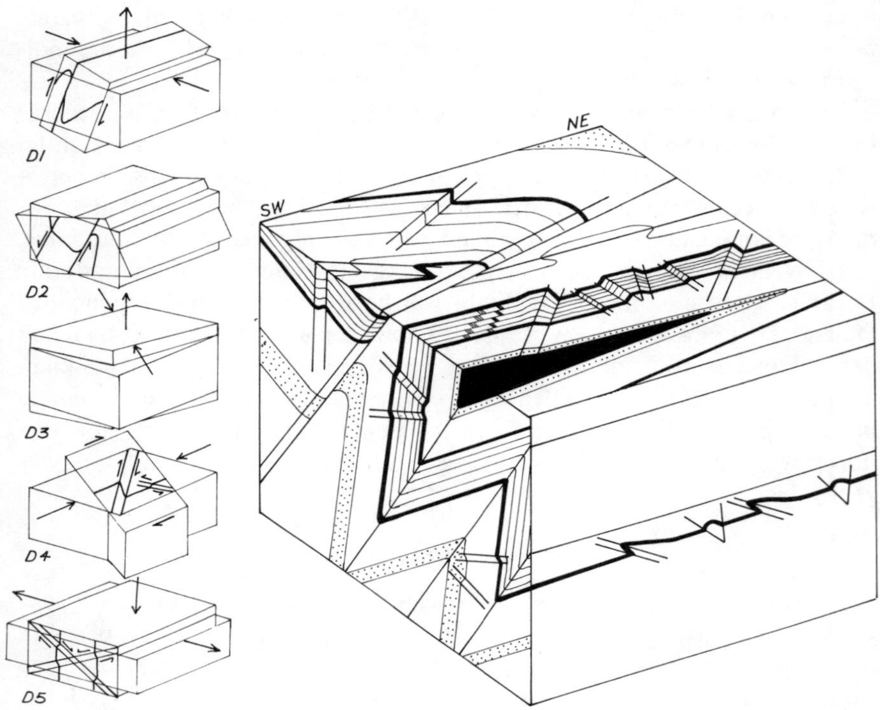

Figure 21. Idealized strain geometry and history of the Arisaig area.

in the northern limb of the Arisaig Syncline between the axial trace of the Smith Brook Fold and the major, D_2, Arisaig Slide downthrowing to the northwest. D_3 deformation was apparently localized to the major D_2 fold limb and consisted of a north-south shortening with probably vertical extension. Open superposed folds possess a near-vertical axial-plane fracture cleavage.

The pattern of D_4 deformation may be explained by a northeast-southwest shortening. A conjugate pattern of steeply plunging Z and S folds on steep D_1 fold limbs developed a monoclinic symmetry, dominated by Z folds, due to a northeast-southwest dextral shear component in foliation. No unique extension direction can be recognized for the D_4 strain regime, because, on steeply dipping D_1 fold limbs, extension was northwest-southeast and on greatly dipping D_1 and D_2 fold limbs was essentially vertical.

D_5 deformation resulted in the local production of a conjugate set of gently inclined kink bands. Shortening was vertical and extension was in a northwest-southeast direction.

The precise stratigraphic age of each of the movement phases is uncertain, except that they are clearly post-Lower Devonian. At least D_1 and D_2 are pre-Carboniferous, as Windsor sedimentary rocks rest unconformably on the major D_2 fold limb in the southwestern part of the area. The precise role played by the Hollow Fault zone during successive movement phases is uncertain; but elsewhere in Nova Scotia, Fyson (1964) has demonstrated variable and complex alternations of dip-slip and strike-slip movements along the proposed southwesterly extension of the fault zone. Dextral strike-slip movement on the Hollow Fault zone probably accompanied the D_4 phase and dip-slip movements, and with downthrow to the north may have accompanied D_5 movements.

Geological History

The geological history of the Arisaig region begins with the poorly dated strata of the Browns Mountain Group. Based on fragmentary inarticulate brachiopods, the Browns Mountain has been tentatively dated as Lower Ordovician. The Browns Mountain beds were tightly folded and intruded by a variety of granitic rocks prior to Silurian time. At School Brook Cove, Cape George (Boucot and others, 1959) fossiliferous Middle or Upper Ordovician quartzites occur, which are also intruded by granitic rocks. Unfortunately, the relations of the fossiliferous rocks at School Brook Cove to the adjacent Browns Mountain Group are not established.

BEARS BROOK VOLCANIC GROUP

Dunn Point Formation

The Bears Brook Volcanic Group was formed under predominantly terrestrial conditions, probably during the Upper Ordovician. Subaerial conditions are proved by the complete lateritic soil profiles and the welded tuffs, and it seems likely that the lavas and sediments were emplaced on a basement consisting of the Browns Mountain Group intruded by monzonite stocks. During Dunn Point Formation time, andesites were extruded and flowed southward from a hinterland to the north in a series of sheets and tongues to build up a tableland. The lateritic soil profiles developed on individual andesite flows give information on the time intervals between flows and on the climate of the period. Thickness of laterite profiles is a function of the time interval between flows, and the absence of a laterite on a flow indicates rapid extrusion of successive flows. The climatic factors indicated by the lateritic soils may be summarized as follows:

1. Lateritic soils are developed at the present day under tropical and subtropical conditions. Thick unconsolidated red laterites are common in tropical regions where a rain forest cover is developed and partially bleached indurated laterites are characteristic of subtropical regions.

2. Lateritic soils are the result of deep and complete hydrolysis of silicates to a clay-mineral complex, usually with the production of kaolinite, gibbsite, and occasionally nontronite. Lateritization of silicic rocks involves the liberation of large amounts of silicic acid as a gel which may give rise to concretions or effect a strong lithification. Intermediate and mafic rocks

yield considerable amounts of iron which in the oxidized state imparts a red coloration to laterites.

3. Lateritization is not solely developed in humid regions with a tropical rain forest cover, and indeed lithified and indurated lateries may only develop where the soil is dried at some stage and deposition of silica and/or carbonate occurs during evaporation. Thus there is no reason why laterites should not develop during the monsoon season in arid regions. A short period of intense deep lateritization during the wet season would be followed by a period of deep drying out during the dry season when the products of deep hydrolysis would be cemented by silica and perdigon nodules.

4. The Bears Brook laterites were probably developed under a seasonal climate for a number of reasons. First, intense seasonal rainfall is indicated by the cutting of the deep wadi channels in the andesite tableland and the slumping of laterites into these channels. Second, no dense vegetation cover could have existed in Upper Ordovician times to bind together an unconsolidated mass of wet laterite developed under humid conditions. Erhart (1956) has assembled evidence which he suggests indicates a land flora of pteridophytes in pre-Devonian times, but no plant fragments have been seen in the Bears Brook laterites. Thus the explanation of the development and preservation of the laterites must be sought elsewhere. It seems a reasonable assumption, that, if the laterites were entirely unconsolidated, they would be largely stripped off by wind action if no vegetation cover existed. Therefore, the writer prefers the explanation that the laterites were preserved on account of their strong induration developed during the dry season in an arid climate.

5. The red laterites occasionally show white reduction spots, which may be evidence of the inclusion of fine organic material. No mechanical breakdown of the laterites has yet been undertaken for the recovery of possible spores, but it seems possible that some protection of the lateritic soils may have been afforded by a carpetlike Thallophyte cover (possibly lichens and blue-green algae). The small pockets of finely bedded laterite (presumably wind-blown) support the notion of an irregular, possibly clumplike, organic cover with wind excavation and deposition in unprotected regions between clumps.

6. The slumped laterites and continuous horizons of bedded wind-blown laterite indicate that although induration and a possible Thallophyte cover afforded only partial protection for the laterite, unconsolidated wet laterite was slumped during the wet season and unconsolidated dry laterite was wind transported during the dry season.

During the later phases of Dunn Point Formation times, when the extrusion of andesites was waning, lateritization attained a firm grip on the landscape, and rhyolites and ignimbrites (ra, ia, and rc, on Fig. 3) were emplaced. The nuée ardente responsible for the ignimbrite "ia" flowed southward down a wadi channel, possibly carrying large blocks of Browns Mountain quartzite. The association of rhyolite extrusion with faulting is demonstrated by the relation of the Doctors Brook Fault with the lower rhyolite (Fig. 3; Pl. 3, secs. 9 and 10). At the end of Dunn Point Formation times, the landscape consisted of a moderately even surface covered by lateritic soils and slumped and wind-blown laterite.

McGillivray Brook Formation

During the emplacement of the McGillivray Brook Formation, the landscape was converted south of the hinterland into a barren uniform surface composed of blanketlike sheets of ignimbrite, with occasional thick prisms of extrusive rhyolite locally forming limited areas of slightly more rugged topography. That laterite soils were still exposed to the north is evidenced by the laterite/ash slumpsheet between "ib" and "ic" and by the windblown bedded laterite above "id" (Fig. 3, secs. 9 and 4).

Malignant Cove Formation

The Malignant Cove Formation is unfossiliferous, but its clastic constituents suggest that it was derived from the erosion of the Bears Brook volcanic rocks, the Browns Mountain Group, the granitic rocks that intrude the Browns Mountain Group, and the Ordovician rocks at Cape George.

ARISAIG GROUP

From Beechhill Cove time (lower Llandovery) to the middle of Stonehouse time (middle Downtonian), currents flowed toward the southwest (Table 6; Fig. 14A). During upper Stonehouse (upper Downtonian) and Knoydart (upper Downtonian and Dittonian) time, flow was toward the northwest (Table 6; Fig. 14B).

Beechhill Cove Formation

Deposition of the basal lower Llandovery conglomerate in stream channels was followed by marine transgression.

The subangular nature of the quartz in the Beechhill Cove siltstones suggests transportation over only moderate distances. A very shallow marine environment is indicated by the brachiopods, which include lingulids in growth position, and by channels, scour depressions, and oscillation ripple marks. However, the 20 percent of matrix material in the siltstones shows that little winnowing of the sediments took place before burial.

The well-sorted silty sediments suggest derivation from a low-lying land mass or deposition by currents too gentle to carry coarse particles.

Ten miles southwest of Arisaig, the Beechhill Cove Formation in northeast Pictou County is similar in lithology and fauna to the beds at Arisaig (Maehl, 1961). But in southeast Pictou County (30 mi southwest of Arisaig), the upper member of the Glencoe Brook Formation, of the same lower Llandovery age, consists of coarse clastics, suggesting deposition closer to the source area; the Pictou County source area may also have had higher relief than that near Arisaig. Both the upper member of the Glencoe Brook Formation and the Beechhill Cove Formation contain *Cryptothyrella;* this is seen in other areas to indicate proximity to land; for example, in the Brassfield Formation, there is a richer *Cryptothyrella* Community from northern Bath County, Kentucky, north to Adams County, Ohio, than southward to Lincoln County, Kentucky (Rexroad and others, 1965).

The evidence from Pictou County thus suggests that land lay to the south or southeast in lower Llandovery times; it may also have existed

TABLE 6. Directions of Flow of Depositing Currents

Directions of flow to nearest 15° — From	Beechhill Cove Fm.	Ross Brook Fm.	French River, Doctors Brook, McAdam Brook Fms.	Moydart Fm.	Basal Stonehouse Fm.	Subtotal	Upper Stonehouse Fm.	Knoydart Fm.	Subtotal
N. to S.	1	indicators rare—no data	4	5	2	12	1	2	3
N. 30° E. to S. 30° W.	20		23	15	12	70	8	1	9
N. 60° E. to S. 60° W.	14		13	7	12	46	3	3	6
E. to W.	2		2	2	4	10	2	1	3
S. 60° E. to N. 60° W.	1		—	1	1	3	7	11	18
S. 30° E. to N. 30° W.	—		—	5	1	6	18	16	34
S. to N.	5		2	5	1	13	6	4	10
S. 30° W. to N. 30° E.	3		—	3	7	13	3	4	7
S. 60° W. to N. 60° E.	3		3	2	2	10	1	—	1
W. to E.	1		2	2	4	9	—	2	2
N. 60° W. to S. 60° E.	—		1	2	4	7	1	2	3
N. 30° W. to S. 30° E.	—		2	1	—	3	—	—	—
Totals	50		52	50	50	202	50	46	96

Current directions were determined mostly by measuring cross-laminations, less commonly by trend of current ripples, channel axes, and scour marks.

to the northwest. The direction of the land supplying the Beechhill Cove sediments is not certain, for the indications show depositing currents from the northeast; if these are parallel to the shore, the land could have been to the northwest or southeast; if they were offshore currents, the land would have lain to the northeast. All that can be said definitely is that the source area was not west or southwest of Arisaig.

Ross Brook Formation

Late in lower Llandovery time, an abrupt change in conditions, perhaps associated with a rise in sea level and a lowering of the source area, is reflected in the sharp basal contact of the Ross Brook Formation. The black graptolitic shales at the base of the lower member are followed by silty shales interbedded with 13 thin tuff layers, some of which contain brachiopods.

The paleogeography changed little until after middle Llandovery time, when siltstones were deposited in the upper parts of the middle member of the Ross Brook Formation. These become progressively more abundant upward in the upper member. The siltstones are associated with *Eocoelia,* a brachiopod which belongs to a marine fossil community only slightly farther offshore than the lingulids seen in the Beechhill Cove Formation (Ziegler, 1965). Scours and oscillation ripples in the upper member confirm the *Eocoelia* evidence for shallow water continuing throughout the upper Llandovery, even though graptolites are present in some beds, and none of the beds are coarser than silt. The increase in siltstones may be due to more rapid erosion in the source area or minor uplift of the source. The presence of graptolites in the lower member of the formation may be associated with periodic reducing conditions in a shallow-water environment unsuitable for brachiopods. Indicators of current directions are rare and difficult to read.

In northeast Pictou County, the beds equivalent to the lower member of the Ross Brook Formation are calcareous mudstones, but the middle and upper members are similar lithologically to the Arisaig beds. In southeast Pictou County, Maehl (1961) correlates the Ross Brook Formation with the shales, slates, and sandstones of the Kerrowgare Formation, which were perhaps laid down in a similar shallow marine environment.

French River Formation

A sudden increase in coarse silt sedimentation coincides very closely with the start of Wenlock times. As in all the Silurian at Arisaig, the grain size is never coarse and originally there was as much as 35 percent matrix (now altered to sericite and chlorite) in most of the arenaceous rocks. The angularity of the quartz and feldspar indicates compositional immaturity; the source of the sediments was therefore most probably an area of igneous or metamorphic rocks.

A 2.5-ft bed of oölitic iron ore is distinct from the siltstones in containing rounded quartz grains. It is probable that like other Silurian iron ores (for example, the Alabama ore; Sheldon, 1964), this bed was formed near the shore; the rounding of the silt particles was probably carried out in the surf zone (Kuenen, 1959, 1964). Judging by the fauna, which is similar to that in the rest of the French River Formation, there was no very radical change in the environment during the formation of the iron ore.

The iron ore is only present in the southeast of the Arisaig area. A similar bed has been recorded in northeast Pictou County, but it too appears to be of limited lateral extent, as it is not present in all the sections of the French River Formation (Maehl, 1961, p. 55). A similar fossiliferous iron ore appears at one locality in the Kerrowgare Formation of southeast Pictou County (Maehl, 1961, p. 87). These iron ores are all of Wenlock age, but it is not possible to ascertain whether they are strictly contemporaneous.

Doctors Brook Formation

In later Wenlock times, shallow-water marine conditions continued. More shale beds are present in the Doctors Brook Formation than in the French River Formation, and the lowest limestones (coquinites) of the Arisaig Group appear in later Wenlock beds. Many channels and scours show that the water remained shallow.

Phosphatic fragments (0.5 to 2.0 cm) are common in some beds; a few of these are inarticulate brachiopods, but most appear to be inorganic. They probably indicate minor pauses in sedimentation. Currents came predominantly from the northeast, as in the Beechhill Cove Formation.

McAdam Brook Formation

Along Arisaig Brook and Ross Brook, the base of this formation is gradational and is marked by a progressive reduction in siltstone beds and by inclusion of darker shales above. On the shores (stations 51 and 52), the basal black shales of the McAdam Brook Formation rest sharply on a minutely scoured surface (scours ⅛ in. deep) of the Doctors Brook Formation. An over-all decrease in the sediment flow from the source area seems to have been coupled with local uplift in one area (the present shore) while sedimentation was continuous a mile or so to the east (Arisaig Brook). The lowest 50 ft consists of black micaceous shales; early Ludlow graptolites are present in one bed. Epifaunal and infaunal forms are present throughout the formation.

Some thin siltstones appear in the upper two-thirds of the lower member, and become progressively thicker and more abundant toward the top of the formation. Most were deposited in scours and some are cross-laminated and show current ripples. The McAdam Brook Formation may represent progressively shallower marine environments, or may have resulted from increasing flow of sediment from a rising source area in a constantly shallow sea; this latter alternative is supported by the continuous presence of brachiopods associated with shallow marine environments.

With current directions from the northeast, and deeper water conditions in the south (Pictou County) and southwest (Annapolis Valley), it seems probable that the Arisaig Ludlow sediments were derived from the northeast or northwest.

Moydart Formation

The Ludlow deposits continue with massive coarse siltstones and fissile shales which make up the green member of this formation. The silts and clays were most probably deposited from suspension; but periodic bottom currents are indicated by the cross-laminated beds, the scours filled with shelly arenaceous limestones, and bedding planes marked by interference and oscillation ripples. As for older formations, current directions appear to be mainly from the northeast, and, as for the McAdam Brook Formation, the source of the sediments is thought to be from the northeast or the northwest.

Toward the end of the Ludlow deposition, sedimentation rates increased compared with subsidence, and the nonmarine red member of the Moydart Formation was laid down. Dineley (1963) compares the red marls, siltstones, and nodular limestones of this member with similar beds in the Knoydart Formation and in the British Lower Old Red Sandstone. The nodular limestones are indistinguishable from beds in the "Psammosteus" limestones of the Ditton Series of Britain; they are probably the result of penecontemporaneous crystallization of carbonate in mud, forming just below the land surface like the Kunkhar limestones of India. The interbedded red marls and siltstones are probably chiefly fluviatile like similar Dittonian beds in Britain (Allen and Tarlo, 1963).

Stonehouse Formation

In Pridoli and lower Gedinnian times, there was a return to the marine environment that was present before the deposition of the red Moydart. The Stonehouse beds contain cross-laminated channel deposits, oscillation and interference ripples, and shelly limestones laid down in scours, all of which point to very shallow marine conditions.

In the top 100 ft or so, there is a marked reduction in the variety of marine fossils, perhaps signifying extreme shallowing before the onset of the nonmarine conditions of the overlying Knoydart Formation.

The lower part of the Stonehouse had a majority of its depositing currents from the northeast (as in earlier marine formations at Arisaig), but the upper part had currents mainly from the southeast. The probable source areas are discussed below in the chapter on regional relations.

Knoydart Formation

The red clastics of the Knoydart Formation are interpreted as having accumulated in a Lower Devonian coastal floodplain region. As in the upper part of the Stonehouse, the current directions were mainly from the southeast, but contemporaneous marine beds in the Annapolis Valley suggest that the sediments were derived from the north or northwest.

The remarkable similarity of the Knoydart Formation to the Lower Old Red Sandstone of Britain is paralleled by its equivalence in age. Boucot (1961) and others have noted striking affinities between the lower Paleozoic faunas of the northern Appalachians and those of Europe. The vertebrates in the Knoydart Formation continue this resemblance despite the fact that they are not marine animals. E. I. White (1957) has commented upon the problem of their distribution over the North Atlantic area and suggests that the pteraspids at least may have spawned in the coastal marine waters and spent an early or larval stage of their lives in the sea where they could migrate over wide distances. In the regrouping of the northern continental masses currently popular with the proponents of continental drift, the distances separating the Nova Scotian and British Paleozoic outcrop areas are drastically reduced and the possibility of a more or less continuous Ludlow/Downton–Stonehouse and Downton/Ditton–Knoydart facies belt is attractive. Allen and Tarlo (1963) have proposed a table showing the change of facies seen in the Ludlovian, Downtonian, and Dittonian beds of the Welsh Borderland. This can be closely compared with what is seen in Nova Scotia (Table 7).

One may possibly envisage a coastal floodplain region in which the "Old Red" sediments accumulated, extending across an area now divided between Britain and Nova Scotia. Truly marine conditions persisted in the western portion of this region until later than was the case in the east, and the details of the swing to nonmarine red-bed sedimentation were somewhat different. Nevertheless, there was a continuity of facies belts and ecological zones that persisted throughout the deposition of the Upper Silurian and lowest Devonian as seen in these now widely separated areas.

TABLE 7. Lithology and Interpretation of the Knoydart Formation and the Old Red Sandstone

England and Wales: Age	England and Wales: Formation	Lithology and Paleontology	Interpretation	Nova Scotia: Formation	Nova Scotia: Age	
DITTONIAN	Ditton Beds	Thick red siltstones; with concretionary limestones. Thin red or green sandstones; cross-bedded, current lineation, scour-and-fill, suncracks, current ripples, channels. Conglomerates on erosional surfaces. Animal burrows. Plants.	*Chiefly fluviatile*: fans of sand spread **by strong rivers after crevassing on** floodplain with temporary silty lakes and mud-flats. Brief invasions giving brackish silty and sandy lagoons. Floods.	KNOYDART	Dittonian	
DOWNTONIAN	Downton Beds	Greater part of succession; thick red or green siltstones, often laminated. Thin green or red micaceous sandstones; cross-bedding, flat-bedding, current ripples Molluscs. Abundant animal burrows and *Lingula*.	*Subtidal and intertidal:* sands of subtidal shoals and lower intertidal flats. Silts of upper intertidal flats. Strong waves and currents. Brackish and turbid water.	absent		PRIDOLI AND LOWER GEDDINIAN
		At base: Yellow sandstones above; scour-and-fill, cross-bedding, flat-bedding, current ripples. Bone beds; shells, phosphatic debris. Green siltstones below: current ripples, lamination. *Lingula* with abundant molluscs.	*Nearshore to beach:* silts formed offshore within wavebase overlain by transgressive sand shoals and beaches. Strong waves. Turbid, often brackish water.	absent	DOWNTONIAN	
LUDLOVIAN	Ludlow Beds	Dominantly green siltstones and shales: current ripples, lamination. Thin sandstones; flat-bedding, current ripples. Occasional discordant bone beds. Limestones; shelly, some authochthonous, others detrital, conglomeratic, cross-bedded. Abundant corals, brachiopods, and molluscs.	*Chiefly open sea:* Clays, silts, sands and limestones formed offshore chiefly within wavebase. Clear sea, often turbid, sometimes brackish. Periodic strand retreat-advance.	STONEHOUSE		

Lochaber Area

During the summer of 1959, a geological reconnaissance of the Lochaber area in Antigonish County (Fig. 1) was made by several of the students at the Massachusetts Institute of Technology summer camp under Boucot's supervision. Although the structural and stratigraphic results are less conclusive then those from nearby areas, this location is geologically strategic, and it is appropriate to list some of the facts observed.

1. A suite of sedimentary, volcanic, and intrusive rock types lying to the west of Lochaber (Pl. 9) is similar to those assigned elsewhere in Antigonish and Pictou Counties to the Browns Mountain Group.

2. To the east of the rocks assigned to the Browns Mountain Group (Pl. 10), there is a sequence of greenish siltstones and slates, together with a scattering of shelly limestones, which can best be referred to as "Arisaig Group undifferentiated." These beds have provided enough diagnostic fossils to demonstrate that they include representatives of the following: French River Formation, McAdam Brook Formation?, Moydart Formation (green member only), and the Stonehouse Formation.

Table 8 details the brachiopods found in these strata.

3. East and southeast of the beds assigned to the Arisaig Group, a sequence of siltstones and fine-grained sandstones that are predominantly reddish are assigned to the lithologically similar Knoydart Formation. The contact with the underlying Arisaig Group is gradational on Cameron Brook (Pl. 10) as well as on the two unnamed brooks to the north of it. The gradational contact consists of an interbedding over a stratigraphic interval of several hundred feet of greenish "Arisaig type" strata with reddish "Knoydart type" strata. Unfortunately, no fossil fish have been recovered from the Knoydart Formation in the Lochaber area, although Dineley has carefully examined the section on McNab Brook, the most complete section of the Knoydart in this area.

4. To the north of Lochaber itself, there are numerous exposures of both the Horton Formation and the Windsor Limestone, both of Carboniferous age. In those places where the red beds of the Horton Formation are heavily cleaved, it is very difficult to differentiate them from the very similar beds of the Knoydart Formation.

TABLE 8. Brachiopods Identified from the Arisaig Group in the Lochaber Area

French River Formation (D. Benson locality B 079B2963)	
Meristina billingsi	
McAdam Brook Formation? locality USNM 11183	Plate 11
Protochonetes novascoticus	
Isorthis sp.	
Howellella moydartensis	
Moydart Formation locality USNM 11185	Plate 9
Eatonioides lamellornatus locality USNM 10827	Plate 9
Protochonetes novascoticus	
Eatonioides lamellornatus	
Stonehouse Formation locality USNM 11182	Plate 11
Protochonetes novascoticus	
"Camarotoechia" planorugosa	
Delthyris rugaecosta locality USNM 11184	Plate 10
Mclearnites mertoni	
Protochonetes novascoticus locality USNM 11186	Plate 9
Isorthis fornicatimcurvata	

5. It is likely that more detailed mapping would provide a sound lithologic basis for subdividing the Arisaig Group in the Lochaber area, and for easily distinguishing between the Knoydart and Horton Formations. The Arisaig Group of the Lochaber area has much more in common with that of the Arisaig area and northeastern Pictou County, than with southeastern Pictou County, and is assigned to the Arisaig Belt.

Lower Paleozoic Rocks of the Cape George Area

INTRODUCTION

This section includes a compilation and reinterpretation of previous reports on the geology of northern Cape George (Pl. 12) and a detailed account of the geology of School Brook and Cormorant Cliff Coves (Pls. 13, 14).

The tip of the Cape is bounded on the east by George Bay and on the north and northwest by Northumberland Strait. The southern limit of the mapped area is a line from Skidmore Brook on the eastern shore to Horseshoe Brook (Pl. 12) on the western shore. The area includes about 16 sq mi and the southern border is 18 mi north of the town of Antigonish on the Cape George (Harbour) Road. The Cape George Road follows the shoreline around the Cape from the Marsh Cove on the east to Livingstone Cove on the west. There is almost continuous outcrop along the shoreline. Wilkie Brook [It is named Wilkie Brook on a Canadian Geological Survey Map of 1893 (Nova Scotia Sheet No. 33). A more recent map of the Canadian National Topographic Series (1953, Sheet No. 11 F/13 West Half) calls it Cape George Brook.] in the southeast part of the area and Livingstone (the 1893 Geological Survey Map calls it Livingstone Brook; the same brook is named Gillis Brook on the 1953 National Topographic Map) and McNeil's Brook (J. R. Griffin states that McNeil's Brook is a tributary of Livingstone Brook) in the western part also provide numerous exposures. There are scattered outcrops along many of the brooks and on the hills of the area.

The first part of the section is a compilation and reinterpretation of maps and reports by Hugh Fletcher (1887); Paul D. Kaminsky (1953); James R. Kramer and William C. Phinney (1953); and field camp notes and maps prepared by George Erlanger and Viet Howard (1959); together with additional information gathered in 1959 by the writers.

The two coves described in the second part of the section are at the northern tip of Cape George. The eastern cove, School Brook Cove, is

1,500 ft long and 150 to 250 ft wide. Cormorant Cliff Cove is ⅓ mi west of School Brook Cove and is 1,000 ft long and 150 to 200 ft wide. The two coves are situated in fault zones that strike east-west and contain blocks of Ordovician, Silurian, and Devonian rocks. These blocks are up to 50 ft across and form a mega-breccia. The tip of the Cape, from north of Livingstone Cove to Ballantyne Cove, consists of Horton conglomerate with the exception of the mega-breccia at School Brook and Cormorant Cliff Coves.

A restudy of the Arisaig Group in Antigonish County, Nova Scotia, was started by Boucot during the summer of 1958. In connection with this work, he and Nathaniel McLain Sage visited the tip of Cape George to look at the two small areas mapped by Hugh Fletcher (1887) as Silurian from which Fletcher collected fossils subsequently described by Ami (1895). During the 1958 visit, it was observed that the area included strata lithologically similar to the "Red Band" of the upper Moydart Formation and the underlying main body of the Moydart. In addition, a number of quartzite outcrops were observed that are not like any rocks present elsewhere in Antigonish County. Some of the quartzite outcrops contain fragmentary fossils, and others contain abundant *Scolithus*-like structures. During the summer of 1958, Raymond Fletcher spent 3 weeks making a 50 ft/1 in. map of these two areas of Silurian rocks. Additional fossils that he found in some of the quartzites were identified by Boucot as Middle or Late Ordovician in age (Boucot and others, 1959, p. 1572). During the summer of 1959, Griffin spent 4 weeks checking Raymond Fletcher's results. Additional structural information was obtained, and larger fossil collections were made, including fossils from the quartzites containing *Scolithus*-like tubes.

We are indebted to Murray Copeland, Geological Survey of Canada, Ottawa, who identified the ostracodes from the Stonehouse Formation, and to Tor Ørvig, Swedish Natural History Museum, Stockholm, who identified and dated vertebrate fragments from quartzites of the Moydart Formation.

DESCRIPTION OF ROCK UNITS

Diorite

In the south-central section of the mapped area (Pl. 12) there is a large diorite intrusive mass.

Lithology. The diorite is light to dark green and on weathered surfaces, is characteristically white. The rock is porphyritic, and the groundmass is fine grained, hard, and highly fractured. Many of the elongate grains have a common orientation. The following information was obtained from thin section by Kramer and Phinney (1953, p. 15):

Composition. "Plagioclase (An_{20}) 50 percent, chlorite (pennine) 20 percent, hornblende 5 percent, epidote 15 percent, sericite 5 percent, opaque minerals 5 percent, and accessory appatite."

Fabric. "Holocrystalline plagioclase occurs as euhedral-subhedral phenocrysts and also as part of the fine-grained groundmass, some plagioclase crystals are fractured and granulated. Chorlite forms around the hornblende. Epidote occurs in anhedral to euhedral form throughout the thin sections. Sericite occurs rimming the plagioclase. Opaque minerals consist mostly of subhedral magnetite."

Age. Diorite pebbles similar to that of the diorite outcrops occur in a conglomerate of the pre-Silurian Browns Mountain Group on Cape George which suggests that the diorite is the oldest rock exposed on the Cape.

Browns Mountain Group

Lithology. On Cape George, the Browns Mountain Group contains siltstone, graywacke, conglomerate, and marble.

The siltstone is gray and black, fine-grained, metamorphosed, and quartzose. It is well bedded, compact, cleaved, strongly indurated, and considerably contorted and fractured. The weathered surface is smooth, medium to dark gray, and porcelainlike in appearance. Secondary calcite veins and gash quartz veins cut the siltstone. The siltstone is commonly interbedded with graywacke and contains a small percentage of pyrite, chalcopyrite, and pyrrhotite. The pyrrhotite occurs along fracture planes within the siltstone, as parallel tabular grains.

The graywacke consists of fine-grained, black and gray-green rocks. It is calcareous and cut by abundant quartz veins, but it is not cut by cleavage planes as is the siltstone. The weathered surface of the graywacke is light gray.

Interbedded with the siltstone, there are a few beds of quartzite and gray-green conglomerate containing well-rounded pebbles that range in size from ¼ to 2 in. across. The conglomerate is massive, strongly cemented by quartz, and weathers light gray. The matrix of the conglomerate consists of a dark gray-green, quartzitic siltstone. The conglomerate pebbles consist of 80 percent gray and black quartzitic siltstone and 20 percent fine-grained diorite.

An exposure of red coarsely crystalline marble was found in McNeil's

Brook about one-half mile south of Marsh Road. The actual stratigraphic position of this marble was not determined, but it is tentatively assigned to the Browns Mountain Group.

Thickness. The exact thickness of the unit is not known, because the structure is poorly understood. The Browns Mountain Group has not been extensively mapped and therefore the thickness of the unit, which is based on outcrop area and bedding attitudes, is only a first approximation. A minimum thickness of 1,200 ft can be obtained along McNeil's Brook (Pl. 12). The trend of bedding along the brook is N. 85° E., dip 70° N. A maximum estimate of 3,000 ft has been made from a cross section drawn through the Browns Mountain Group from the upper region of Morar Brook to the southern region of Livingstone Cove (Pl. 12). The strike varies from N. 45° E. to N. 85° W. and the dip from 80° N. to 30° N.

Lower Contact. The Browns Mountain Group was probably deposited on the erosional surface of the diorite. The lower contact of the Browns Mountain Group is not exposed, but exposures of lower Browns Mountain Group are separated by a 200 ft covered interval from outcrops of diorite along McNeil's Brook. The strike of the lower Browns Mountain Group near the diorite is concordant with the mapped outline of the diorite. The Browns Mountain conglomerate contains diorite pebbles similar to the diorite along the coast south of Livingstone Cove.

Age. No diagnostic fossils have been found in the Browns Mountain Group either at Cape George or at any other locality in Nova Scotia. The unit is probably the oldest stratified rock type on Cape George. The Browns Mountain Group is tentatively placed in the Cambrian-Ordovician. The unit is believed to be of pre-Taconic age, as it has been more highly altered than have associated Silurian and Early Devonian strata in the area.

Rhyolite

In the upper part of Wilkie Brook (Pl. 12) and on Harding's Mountain to the south (south of area shown on Pl. 12), there are a series of thick rhyolite beds. The rhyolites are highly fractured and deformed flows, agglomerates, and tuffs.

Lithology. The fresh rhyolites are red to yellow-brown. The groundmass of the rhyolite is ochre-red, hard, and aphanitic. The weathered groundmass is soft, crumbly, and cream-colored. The rock contains phenocrysts of orthoclase, quartz, and green spherulites of unknown composition. The phenocrysts weather out and give the porphyritic rhyolite a vesicular appearance. The flows contain bands of light-pink chalcedony that are 4 to

6 in. thick. In the upper Wilkie Brook area, one of the rhyolite outcrops exhibits columnar jointing.

The following data were obtained by Kramer and Phinney (1953, p. 23) for a rhyolite porphyry from Wilkie Brook: the rock is reddish brown, hard, and porphyritic. Medium-sized, euhedral phenocrysts of quartz (30 percent) and orthoclase (20 percent) occur in a fine-grained matrix (50 percent) consisting of quartz, orthoclase, and hematite. Throughout the rock, there are patches of hematite and lighter-colored bands of chalcedony.

Thin-section analysis by Kramer and Phinney (1953, p. 24) of an agglomerate from upper Wilkie Brook, which they describe as dirty white, flinty, well cemented, and resistant to weathering, provides the following information. The thin section contains subhedral phenocrysts of quartz (15 percent), a minor amount of phenocrysts of orthoclase (<1 percent), a devitrified groundmass (60 percent), chalcedony (10 percent), and an undetermined green spheroidal mineral (15 percent). The chalcedony fills cracks and cavities and the green spheroidal mineral fills some cracks and also may surround the orthoclase phenocrysts.

Both linear and curved fractures are found in the devitrified groundmass, and there are flow lines in the groundmass which curve around the quartz phenocrysts. The subhedral quartz grains have relict euhedral outlines.

Interbedded in the rhyolite are green tuff beds with phenocrysts of quartz and feldspar, and spherulites.

Thickness. The exposed thickness of the rhyolite is 600 to 800 ft.

Lower and Upper Contact. The lower contact is not exposed in the area, and neither the type of contact nor the nature of the underlying unit is known. The rhyolite is in fault contact with the Browns Mountain Group to the west.

Age. The age and stratigraphic position of the rhyolite is unknown. The unit is either part of the Browns Mountain Group (Cambrian-Ordovician), or it is correlative with the volcanics of the Arisaig area which immediately underlie the Lower Silurian Beechhill Cove Formation.

The unit has been tentatively placed stratigraphically above the Browns Mountain Group and below the Silurian age rocks.

Ordovician to Devonian

The following units are described in the section, "The lower Paleozoic Rocks of the Cape George Area, Antigonish County" under "Geology of School Brook and Cormorant Cliff Coves."

Middle or Upper Ordovician. Hornfels and quartzite.
Hornblende Diorite. (in School Brook Cove).
Silurian. *Scolithus* (worm tube) quartzite and Moydart Formation.
Silurian and Devonian. Stonehouse Formation.
Devonian. Knoydart Formation.

Carboniferous

Horton Formation (conglomerate and sandstone). The tip of Cape George north of Livingstone and Ballantyne Coves consists entirely of Horton Formation, except for School Brook Cove and Cormorant Cliff Cove which contain Ordovician, Silurian, and Devonian rocks in the fault zones. The Horton Formation can be divided into an upper sandstone and an underlying conglomerate unit. In general, the conglomerate can be further broken down into three parts: a lower basal conglomerate, a sandy conglomerate, and a silty conglomerate. The changes from basal conglomerate to sandy and then to silty conglomerate are not obvious and sharp, but are gradational. Thus, the conglomerate is a continuously gradational unit with pebbles and matrix becoming finer upward.

Basal Conglomerate. The basal conglomerate of the Horton Formation is exposed in two areas: at the west end of School Brook Cove (Pl. 13) resting unconformably on top of Silurian and Devonian strata, and south of Livingstone Cove where it rests unconformably on Browns Mountain conglomerate.

Lithology. The conglomerate at School Brook Cove is dark gray-green near its base on both weathered and fresh surfaces, and red in the upper portion of the unit. The reason for this is the color of the matrix, which is dark gray-green near the base but which becomes reddish in the upper portion of the unit due to hematite in the argillaceous material. In the conglomerate, there are large scours and fills, large-scale cross-bedding, shingling, and graded beds.

The clasts of the conglomerate are angular to subangular and have their long directions parallel to bedding. They range in size from pebbles ½ in. in length to boulders 5 ft in greatest dimension. The average size of the clasts throughout the lower portion of the unit is 10 to 12 in., and they become smaller toward the top of the unit.

The clasts consist of granitic rocks (35 percent), hard argillite (15 percent), diorite and greenstone (10 percent), limestone (5 percent), red siltstone and sandstone (10 percent), quartz (5 percent), porphyritic felsite (5 percent), and quartzite (10 percent). About 5 percent of the volume of the pebbles consists of crosscutting calcite veinlets. The matrix consists of angular fragments of quartz (35 percent), orthoclase (15 percent), plagioclase (10 percent), siltstone (10 percent), epidote (5 percent), pyroxene and amphibole (5 percent), and coarse calcite (5 percent). The cement and interstitial material (10 percent) consists of silica, calcite, and shale.

The basal conglomerate of the Horton Formation south of Livingstone Cove is colored dark green, but commonly is stained red by hematite. The conglomerate is made up of subangular pebbles and is well cemented by quartz and some calcite. The lower portion of the conglomerate, due to the angularity of the pebbles, resembles a fault breccia 2½ inches in width. The basal 2½ inches consists of gray and black quartzitic siltstones (35 percent), granitic rocks (25 percent), quartz (10 percent), diorite (10 percent), greenstone (5 percent), and red siltstone and sandstone (15 percent).

Thickness. The thickness of the basal conglomerate varies considerably from one locality to another; at School Brook Cove, it is 500 ft thick and at Livingstone Cove it is 200 ft thick.

Lower Contact. The lower contact is exposed at the west side of School Brook Cove where the Horton Formation rests with angular unconformity upon the Devonian Knoydart Formation. The Horton strikes N. 25° E., dips 35° W. and the Knoydart strikes N. 28° E., and dips 50° W.

The lower contact of the Horton Formation is also exposed 1,500 ft south of Livingstone Cove where it rests with angular unconformity on Ordovician conglomerate of the Browns Mountain Group. The Horton here strikes N. 75° W., and dips 70° N., whereas the Browns Mountain strikes N. 30° W., and dips 50° N.

Age. The conglomerates are believed to be of Horton age. A plant fossil *(Lepidodendropsis)* of Horton age was found on the north coast of Cape George, and identified by W. H. Bell, Geological Survey of Canada, but it could not be determined from identification of the fossil whether the conglomerate was of lower, upper, or middle Horton age. The conglomerate is thrust over the younger Windsor Formation at Ballantyne Cove.

Sandy Conglomerate

Lithology. The sandy conglomerate consists of interbedded purple-beige sandstone (10 percent), and purple-beige conglomerate (90 percent).

The conglomerate is massive, red-stained due to hematite, and less resistant to weathering than are the sandstone beds. Conglomerate beds are 3 to 10 in. thick and contain lenses of sandstone. The pebbles, cobbles, and boulders are subrounded to rounded with some angular particles. The largest boulders are 15 in. across.

The conglomerate consists of pebbles and boulders of granitic rocks (30 percent), quartzitic sandstone (10 percent), greenstone (5 percent), limestone (5 percent), porphyritic felsite (10 percent), quartz (10 percent), and matrix (30 percent). The matrix consists of red arkosic sandstone.

The sandstone beds are 1½ in. thick and show cross-bedding. The sandstone is stained red by a small amount of hematite in the cement and is moderately well cemented by calcite and argillaceous material. Its composition is arkosic, containing quartz (60 percent), orthoclase (15 percent), and plagioclase (20 percent) with minor amounts of pyroxene, amphibole, epidote, chlorite, and muscovite.

Thickness. A thickness of 3,000 to 4,000 ft has been estimated from shore sections made north of Ballantyne Cove on the east coast of Cape George, and north of Livingstone Cove on the west coast of Cape George (Pl. 12). This figure is uncertain due to the many faults of unknown relative movement that cut the conglomerate as well as the impossibility of determining the actual upper and lower contacts of the sandy conglomerate because of gradual transition of basal to sandy and from sandy to silty conglomerate.

Lower Contact. The sandy conglomerate, which was not mapped separately from the rest of the Horton Formation has a transitional, indefinite contact with the underlying basal conglomerate.

Age. The unit locally forms the middle portion of the Horton Formation (conglomerate) and is of Mississippian age.

Silty Conglomerate

Lithology. The silty conglomerate is similar to the sandy conglomerate except that the matrix is finer grained and is redder due to hematite.

Interbedded in the silty conglomerate are beds of red siltstone and sandstone. The siltstone's reddish color is also due to hematite. The siltstone is similar to the matrix of the silty conglomerate.

Lower Contact. The lower contact is gradational with the underlying sandy conglomerate.

Thickness. The thickness was not obtained because the upper portion of the unit was not observed. About 500 to 1,000 ft of the silty conglomerate is exposed.

Age. The unit forms the upper portion of the Horton Formation (conglomerate).

Horton Formation (Sandstone)

Lithology. The sandstone is light brown, locally reddish due to hematite staining, micaceous, and well cemented. The weathered sandstone is light gray and crumbly and is in blocky beds 4 in. to 1 ft thick. The sandstone beds contain cross-bedding and concretions 5 to 6 in. long (the concretions consist of dirty-gray sandstone that is harder than the surrounding rock). The grains are well-sorted, subrounded, medium to very coarse sand-sized, and are cemented by secondary calcite. The composition of the sandstone (determined by Kramer and Phinney, 1953, p. 21) is quartz (55 percent), plagioclase (An_{20}) (10 percent), orthoclase and microcline (less than 5 percent), rhyolite fragments (5 percent), muscovite and biotite (15 percent), and opaques and chlorite (10 percent).

Interbedded with the sandstone are red conglomerate and alternating red and gray-green shale. The conglomerate is red due to staining by hematite. It consists of massive beds and contains well-rounded pebbles up to 1 in. long. These pebbles consist of red and green siltstones (25 percent), quartz (15 percent), greenstone (5 percent), granitic rocks (5 percent), red sandstone (20 percent), and quartzite (15 percent). The matrix of the conglomerate consists of light-brown sandstone.

The red and green shale occurs in thin beds ½ to 1 in. thick, which is soft and slightly calcareous. The red and green colors alternate but may occur in irregular patches instead of following bedding. The gray-green shale is less calcareous than the red shale. The shale layers are interbedded with the light-brown sandstone and red conglomerate beds.

Thickness. The unit is 1,000 to 1,500 ft thick.

Lower Contact and Upper Contact. The lower contact is not exposed in the mapped area. The upper contact is an angular unconformity with the Windsor Formation.

Age. The unit is placed in the upper Horton Formation since it unconformably underlies the Windsor Formation and either overlies the conglomerate of the Horton Formation or is a sandy facies of the conglomerate. The sandstone contains numerous unidentified plant fragments.

Windsor Formation

Lithology. The Windsor Formation consists of marine shale, sandstone, algal limestone, and limy conglomerate. The lithology of the Windsor Formation near Wilkie Brook is different from the usual marine Windsor Formation that is found throughout Nova Scotia. The absence of the normal thin limestone beds of the Windsor as well as the presence of algal reef limestones, large cross-beds as much as 8 ft across, mud-cracks, and a considerable amount of conglomerate suggest a nearshore or shoreline deposit.

The lower part of this unit consists of reddish sandstone and conglomerate. The conglomerate contains well-sorted, rounded to subrounded pebbles and a soft, red siltstone matrix. The conglomerate is massive and resistant to weathering. The sandstone consists of well-rounded grains of medium to coarse sand-sized quartz. Scattered throughout the sandstone are pebbles. The lower sandstone near the unconformity contains large cross-beds 4 to 8 ft across.

The conglomerate and sandstone become more calcareous in the upper portions of the units. The sediments are finer, and there are interbedded red siltstone and sandstone. Some of the sandstone beds contain groove casts on their undersides.

Limestone beds appear in the upper portion of the unit. There are beds of algal limestone, oölitic nodular limestone, and a bed of gypsum. Some of the limestone beds are algal reefs. There are few individual reefs, and the largest is only 30 ft thick. The reefs thin or pinch out toward the west. One limestone bed 60 ft thick is yellow, porous, and has numerous pits due to the rapid weathering of the more calcareous parts of the rock. This bed is believed to be correlative with the "Canary" limestone (Unit A_2) of the Windsor Formation (N. McI. Sage, 1959, personal commun.).

Table 9 is a detailed stratigraphic column that was made by Kramer and Phinney. The top of the section is cut by a fault and the Windsor Formation rests with angular unconformity upon sandstone of the Horton Formation.

Thickness. The exposed portion of the Windsor Formation is 1,000 ft thick. An 860-ft thickness is exposed along the shore north of Wilkie Brook.

Lower Contact. The Windsor Formation rests with angular unconformity on sandstone of the Horton Formation.

Age. The unit contains fossils in the limestone beds, but no fossils have been identified. The unit is Mississippian and is correlated with the (A_2) Windsor Formation. The reason for the correlation is the lithologic similarity of yellow pitted limestone to the "Canary" (A_2) limestone of other Windsor Formation sections.

McARAS BROOK FORMATION

Flows and Conglomerates at Ballantyne Cove

Lithology. The unit consists of conglomerate interbedded with mudstone, sandstone, and basalt flows. The sediments are red due to hematite contained in the matrix and cement. The pebbles in the conglomerate have a dark red-brown, shiny, hematite coating.

The conglomerate consists of platy subrounded pebbles and cobble-sized fragments up to 10 in. long. The composition of the pebbles and cobbles, according to Erlanger and Howard (1959, p. 11)[2] is quartzitic siltstone, red siltstone, and red sandstone, with only minor amounts of granitic pebbles. The small percentage of granitic pebbles has been used by Erlanger and Howard for field identification of the unit, as Horton and Browns Mountain conglomerates in the area have a higher percentage of granitic pebbles.

The matrix of the conglomerate consists of poorly cemented clay, silt, and sand-sized material which is stained red by hematite, and which is 20 percent of the volume of the conglomerate. The composition of the matrix is: clay-sized material (70 percent), quartz (20 percent), feldspar (10 percent). The quartz grains are medium sand-sized and well rounded. Clay and calcite cement the rock.

The mudstone is red due to hematite, and occurs in beds 4 to 8 in. thick. The beds contain ripple marks, cross-bedding, and raindrop imprints. The mudstone is soft and weathers rapidly.

The sandstones are medium-grained, red, and are cemented by calcareous silt.

There are five or more basalt flows, ranging from 10 to 70 ft thick. The basalts are dark greenish brown, highly fractured, and fine-grained. The basalts contain contorted lenses of red mudstone that became entrapped in the flows, and they are cut by numerous thin veins of calcite. Vesicles occur throughout the basalt, but are more common and larger near the tops of the flows. The weathered surface of the rock becomes crumbly and the vesicles are more pronounced. Many of the vesicles contain calcite and chlorite fillings.

A detailed stratigraphic section of the McAras Brook Formation at Ballantyne Cove appears in Kramer and Phinney (1953, p. 27) and is reproduced in this paper as Table 10. This section starts at the fault contact between the Windsor Formation sediments and the McAras Brook Formation, and ends at the Livingstone-Ballantyne Cove Fault.

The following is a thin section description of the basalt made by Kramer and Phinney (1953, p. 35): "Plagioclase (An_{25-30}) 55 percent, calcite 20 percent, chlorite 5 percent, undeterminable 20 percent. Fine-grained euhedral plagioclase; opaques are finely disseminated, contains vein-like calcite; chlorite in some of the vesicles."

Thickness. The basalt flow and conglomerates are 600 to 800 ft thick.

Upper and Lower Contact. At both Ballantyne and Marsh Coves, the lower contact is a fault contact, and the upper contact is not exposed. At

---

[2]Erlanger and Howard incorrectly call the unit "Silty Conglomerate Unit (Miss.?)" but state "This unit is tentatively correlated with the McAras Brook Formation (Miss.) of the Arisaig area to the west."

Livingstone Cove, conglomerate that has been correlated with the McAras Brook Formation of Ballantyne Cove lies conformably above the Horton basal conglomerate, and the upper contact is a fault contact.

Age. The basalt flows and the red conglomerate containing platy pebbles are lithologically similar to the flows and conglomerates in the type

TABLE 9. Shore Section, Windsor Formation between Wilkie Brook and Marsh Cove*

NORTH	MARSH COVE
Thickness	*Lithology*
260 ft—Fault—	Covered interval The area is concealed except for a possible outcrop of algal limestone, large weathered conglomerate boulders in the float. A northeast-striking fault is believed to intersect the shore in this interval.
[860 ft]	Stratigraphic thickness from the angular unconformity to the covered interval.
60 ft	Yellow, porous, conglomeratic limestone containing angular shards of an algal limestone. Interbedded with 1 in. arenaceous limestone which has been deformed into small drag folds.
24 ft	Gray-green argillaceous sandstone.
215 ft	Large, poorly sorted, angular conglomerate in red sandy calcareous matrix.
1½ ft	Angular conglomerate with white limy matrix.
60 ft	Conglomerate, poorly sorted with large angular pebbles and a red sandy calcareous matrix. Much lensing.
6 in.	Red limestone, partly oölitic and partly laminar due to algal growth. Prominent mud-cracking at the base.
65 ft	41 sequences of limy red sandstone, 2 to 9 in. thick, and brecciated red limy arenaceous shale, 2 to 9 in. thick. Four 1 to 3 ft algal limestone beds also occur in the sequence.
30 ft	Red and white mottled algal limestone reef rock. The reef pinches out to the northwest, and continues for an unknown extent seaward. Mud-cracks occur in the red shale at the base of the reef.
15 ft	30 sequences of limy red sandstone, 2 to 9 in. thick, and brecciated red limy arenaceous shale, 2 to 9 in. thick. Many of the sandstone beds contain underside groovings.
1 ft	Red and yellow banded limestone, partly recrystallized.
10 ft	Green calcareous brecciated shale.
225 ft	41 sequences of red sandstone, gray shale, green arenaceous shale, and hard, partly recrystallized limestone containing red and green bands. Shale grades from sandstone to conglomerate in an easterly direction. Oscillation ripple marks show an east-west current.
85 ft	Concealed.
4 ft	Conglomerate with angular (8 in.) pebbles of predominantly metamorphic rocks cemented with a white calcareous matrix.

McAras Brook Formation near Arisaig, and therefore the rocks at Ballantyne Cove are correlated with the McAras Brook Formation. No fossils have been found in the conglomerate, and since both the northern and southern contacts are fault contacts, it is not known what unit overlies or underlies the flows and conglomerates.

TABLE 9. (*continued*)

NORTH		MARSH COVE
Thickness		*Lithology*
1½ ft		Calcareous conglomerate with angular pebbles (2 in.) interbedded wih thin calcareous sandstone lenses.
15 ft		14 sequences of red sandstone, gray shale, green arenaceous shale, and hard, partly recrystallized limestone with red and green bands. Limestone nodules distort shale beds above and below.
8 ft		Interbedded fine-grained red conglomerate and red arenaceous limestone with individual beds 6 in. to 1 ft thick.
	1½ in.	Arenaceous red shale.
	1½ in.	Gray-green recrystallized limestone.
2 ft		Arenaceous red shale.
4 ft		Conglomerate, well-cemented with a calcareous red sand matrix; large pebbles (5 in.), containing many sand lenses.
1½ ft		Conglomerate with large angular poorly sorted pebbles in a red calcareous matrix.
3 ft		Calcareous red sandstone.
6 ft		Red conglomerate.
	3 in.	Green calcareous sandstone.
3 ft		Red conglomerate.
	3 in.	Green calcareous sandstone.
	1 in.	Red conglomerate.
	1½ in.	Green calcareous sandstone.
5 ft		Interbedded red conglomerate and sandstone. Conglomerate in 1-ft beds and sandstones in 2-in. beds. The angular pebbles in the conglomerate are ½ in. to 1 in. and are cemented in a sandy matrix. Lowest exposed Windsor Formation. Angular unconformity, 660 ft north of mouth of Wilkie Brook.
660 ft		Horton Formation. Mouth of Wilkie Brook.
SOUTH END OF SECTION		WILKIE BROOK

*After Kramer and Phinney (1953).

TABLE 10. McAras Brook Formation between Ballantyne Cove and Marsh Cove*

NORTH	LIVINGSTONE-BALLANTYNE COVE FAULT
Thickness	*Lithology*
	Horton conglomerate
——Fault——	Livingstone-Ballantyne Cove Fault.
~200 ft	Additional basalt flows and conglomerate which are too faulted to be accurately measured.
....	Mouth of Ballantyne Cove Brook.
40 ft	Coarse red conglomerate.
30 ft	Basalt flow. N. 65° E. 90°.
113 ft	Coarse red conglomerate containing large platy pebbles and having a large number of pebbles with a shiny black oxide coating. N. 85° E., 80° N.
440 ft—Fault?—	Covered interval, possible fault. Kramer and Phinney (1953, p. 27) believe a fault in this interval repeats the section. Erlanger and Howard (1959, p. 12) do not believe the fault in this concealed area repeats the section but that the axis of a syncline occurs here with a north-topping limb to the south, and a south-topping limb to the north.
15 ft	Gray limestone and limy arenaceous green shale. N. 45° E., 45° NW. Rock is brecciated.
65 ft	Coarse red conglomerate.
12 ft	Green and red banded limestone. Approximately twenty 3-in. limestone beds grading to red sandstone in a northeasterly direction along the strike. N. 75° E., 55° N.
35 ft	Coarse red conglomerate. N. 65° E., 40° N.
175 ft	Coarse, red, poorly sorted conglomerate composed of many platy metamorphic pebbles. N. 85° E., 76° N.
70 ft	Coarse red conglomerate. N. 85° E., 45° N.
225 ft	Series of five basalt flows ranging in thickness from 10 to 70 ft. The tops and bottoms of the flows are covered with a conglomerate of basalt pebbles with a red mud matrix due to extrusion on a mud-covered surface.
20 ft	Conglomerate. Coarse, poorly sorted, containing green platy metamorphic and sandstone pebbles in a red sandy matrix. N. 80° E., 79° N.
——Fault——	N. 60° E.
440 ft	Basalt and conglomerate, badly fractured and sheared. N. 60° E., 45° S.
230 ft	Conglomerate, faulted and sheared along vertical surfaces. Many pebbles have a shiny black coating.
15 ft	Gypsum. N. 60° E., 90°. (may be a slice of Windsor Formation within the fault zone.)
260 ft—Fault—	Covered interval. The area is concealed, except for a possible outcrop of algal limestone, weathered conglomerate boulders in the float. A northeast-striking fault is believed to intersect the shore in this interval.
	Windsor Formation.
SOUTH	MARSH COVE

*After Kramer and Phinney (1953, p. 27).

The stratigraphic position of the flows and conglomerates found near Ballantyne Cove is uncertain. There are three possibilities: (1) The unit belongs to the Horton Formation and lies above the conglomerate of the Horton Formation and below the sandstone of the Horton Formation. (2) The unit belongs to the Horton Formation and lies above the sandstone of the Horton Formation and below the angular unconformity with the Windsor Formation. (3) The unit belongs to the Windsor Formation and lies conformably above the Windsor section that is found north of Wilkie Brook. The flows and conglomerates are post-"Canary" (A_2) limestone.

Kaminsky (1953) and Kramer and Phinney (1953) consider the flows and conglomerates at Ballantyne Cove to overlie the Windsor sediments to the south (stratigraphic possibility no. 3). The evidence for this interpretation is that the "Canary" (A_2) limestone of the Windsor Formation is exposed immediately south of the basalt flows and conglomerates and the limestone is the uppermost part of the Windsor observed in the Marsh Cove-Wilkie Brook region of Cape George.

The uncertainty of this stratigraphic position is caused by the presence of a northeasterly striking fault that separates the Windsor Formation from the flows and conglomerates of Ballantyne Cove. The direction and amount of movement along the fault is not known. If the predominant movement was strike-slip, the interpretation of Kaminsky (1953) and Kramer and Phinney (1953) would be acceptable. However, if the fault was a thrust fault and the northwest side, which includes the flows and conglomerates of Ballantyne Cove, was thrust up and over the Windsor Formation, then the rocks of the McAras Brook Formation of Ballantyne Cove would be older than the rocks of the Windsor Formation.

The reddish-brown manganese-stained conglomerates south of Livingstone Cove may be part of the McAras Brook Formation, and if they are, the flows and conglomerates of Ballantyne Cove would thin out toward the west. The evidence for correlation of the conglomerates at Livingstone Cove and Ballantyne Cove, which would assign them to the the McAras Brook Formation, according to Erlanger and Howard (1959, p. 10), is stated by them as follows: "The negligible amount of granite inclusions in the unit is one of the most important methods for field identification; another diagnostic feature is the very dark red-brown, shiny manganese staining of inclusion surfaces and matrix; this staining occurs only in this unit."

Erlanger and Howard place the unit stratigraphically above the basal conglomerate of the Horton Formation and below the light-brown sandstone of the Horton Formation (stratigraphic case 1). Kaminsky (1953) correlates the conglomerate south of Livingstone Cove with the conglomerate of Ballantyne Cove (McAras Brook Formation) on the basis of the presence of an andesitic sill within the conglomerate near Livingstone Cove.

Andesite dikes cut the basal conglomerate of the Horton Formation along the northern shore of the Cape. The andesite dikes are presumed to be related to the McAras Brook Formation flows; therefore, the McAras Brook Formation is younger than the basal conglomerate of the Horton Formation.

Williams (1914, p. 77) states, in reference to the McAras Brook Formation south of Arisaig, "From the apparent conformability of the McAras

TABLE 11. Stratigraphic Section Present on Northern Cape George

	THICKNESS
Mississippian	
Windsor Formation (south of Marsh Cove)	1,000 ft
Horton Formation (sandstone) (north of Ballantyne Cove)	1,000 ft to 1,500 ft
McAras Brook Formation (Horton or Windsor) (flows and conglomerate) (in Ballantyne Cove)	600 ft to 800 ft
Horton Formation (conglomerate)	
Silty conglomerate	500 ft to 1,000 ft
Sandy conglomerate	3,000 ft to 4,000 ft
Basal conglomerate	200 ft to 500 ft
Devonian	
Knoydart Formation (in School Brook and Cormorant Cliff Coves)	min. 50 ft
Silurian-Devonian	
Stonehouse Formation (in School Brook and Cormorant Cliff Coves)	min. 85 ft
Silurian	
Moydart Formation (in School Brook and Cormorant Cliff Coves)	min. 115 ft
Scolithus quartzite (in School Brook and Cormorant Cliff Coves)	min. 245 ft
Hornblende diorite (in School Brook Cove)	
Middle or Upper Ordovician	
Hornfels (in School Brook and Cormorant Cliff Coves)	min. 40 ft
Quartzite (in School Brook and Cormorant Cliff Coves)	min. 30 ft
Cambrian-Ordovician	
Rhyolite	600 ft to 800 ft
Browns Mountain Group	1,200 ft to 3,000 ft
Diorite	

Brook Formation with the overlying limestones of undoubted Windsor age, from the limy condition of much of the formation, and its close resemblance to the strata succeeding the limestones, the writer has concluded that the age of the McAras Brook Formation is essentially that of the Windsor rocks; that is, these strata were deposited in the Mississippian period a short time before the Windsor series." Windsor limestone E_1 (N. Mcl. Sage, 1959, personal commun.) lies conformably above the McAras Brook Formation in the type area.

The interpretation followed in Table 11 is that of stratigraphic case 1.

STRUCTURE OF CAPE GEORGE

Northern Cape George is divided into three distinct geologic areas by two faults. The northern tip of the Cape consists mainly of an east-west–striking homocline of conglomerate of the Horton Formation. The southwestern region consists of Browns Mountain Group, diorite, and Carboniferous conglomerate and flows. The southeastern area consists of sandstone of the Horton Formation, Windsor Formation, and Cambro-Ordovician rhyolite. One fault strikes east-west and extends from Livingstone Cove to Ballantyne Cove. The second fault is largely covered but can be traced by its topographic expression. This second fault intersects the shore at Marsh Cove and the fault trace strikes northeasterly with the older Browns Mountain Group to the northwest and the Horton and Windsor Formations to its southeast.

The Livingstone-Ballantyne Cove Fault and the faults at School Brook Cove and Cormorant Cliff Cove are post-Horton and probably post-Windsor in age. The northeast-striking fault at Marsh Cove is post-Windsor.

Southwestern Area

The southwestern region is bounded to the north by the Livingstone-Ballantyne Cove Fault and to the southeast by a northeasterly striking fault that intersects the shore at Marsh Cove. The area contains Browns Mountain Group, diorite, conglomerate of the Horton Formation, and conglomerate and flows that may be part of the McAras Brook Formation or part of the Horton Formation. The structure of the Browns Mountain Group probably is not as simple as is indicated by the few dip and strike symbols shown on the map (Pl. 12), and further detailed mapping is required. The Browns Mountain Group is affected by intense folding which has produced tight folds and drag folds. In the center of the exposures of Browns Mountain Group, there are outcrops of diorite. This diorite has either been intruded into the Browns Mountain or lies unconformably beneath it. The relation existing between the Browns Mountain Group and the diorite is in question since the contact has not been seen. Attitudes of the Browns Mountain Group are concordant with the diorite contact except along John Joe's Brook, south of the map area (Pl. 12), where one sequence of strata dips toward the diorite instead of away from it, as do most of the beds (Kramer and Phinney, 1953). Bedding in the Browns Mountain Group near the diorite is steep and the beds on the western side of the diorite have near-vertical dips. The steepest dips occur adjacent to the diorite contacts, with the angle decreasing away from the diorite to a maximum of 40° at a distance of several hundred feet.

The conglomerate of the Horton Formation overlies the Browns Mountain Group unconformably. The contact is exposed along the shore ¼ mi south of Livingstone Cove. This conglomerate is overlain conformably by a conglomerate that has been correlated with the McAras Brook Formation.

Conglomerate and basalt flows, which also resemble rocks of the McAras Brook Formation, are found in the vicinity of Ballantyne Cove. They are in fault contact with the conglomerate of the Horton Formation which lies to the north across the Livingstone-Ballantyne Cove Fault. The conglomerate and basalt flows are exposed along the shore south of Ballantyne Cove for a distance of 1,000 ft, where they are in contact with the Windsor Formation along a fault which forms the southern boundary of the southwestern area.

Along the shore south of Ballantyne Cove, the McAras Brook Formation is not exposed for an interval of 440 ft. Kramer and Phinney (1953) suggest that a fault occurs in this interval and repeats the section. However, Erlanger and Howard (1959, p. 12) have observed that the sedimentary and flow features of the conglomerates and flows of the McAras Brook Formation indicate that the beds top to the north on the south of the covered interval, and top to the south on the north of the interval. Thus the beds are not repeated by a fault, but instead they are the two limbs of a tightly folded syncline whose northern limb is overturned. The axis of this syncline strikes east-west and the axial plane of the fold dips north. The tight fold may have been produced by an overthrust of the conglomerate of the Horton Formation which lies to the north.

The basalt flows and conglomerates at Ballantyne Cove are intruded by numerous basalt dikes and are cut by many minor faults. The dikes are very similar in appearance to the flows. When the flows and dikes are faulted a complex structure is produced in which it is very difficult to distinguish between dike rock and flow.

Northern Area

The southern boundary of the northern area is the Livingstone-Ballantyne Cove Fault. The tip of the cape north of Livingstone Cove and Ballantyne Cove consists entirely of the conglomerate of the Horton Formation except for School Brook Cove and Cormorant Cliff Cove which contain Ordovician, Silurian, and Devonian rocks. The Horton Formation has been folded into an east-west–striking syncline in which the beds on the northern limb are vertical and in some places overturned. On the southern limb, the beds dip 40° to 50° to the north. The Horton Formation rests with an angular unconformity on Devonian strata of the Knoydart Formation at School Brook Cove. The Knoydart Formation strikes N. 28° E. and dips 50° W. and the conglomerate of the Horton strikes N. 25° E. and dips 35° W. On the southern limb, the lower contact of the Horton is not seen because the lower part of the unit has been cut out by the Livingstone-Ballantyne Cove Fault.

The conglomerate of the Horton Formation outcropping in the north end of the Cape has been thrust over the McAras Brook Formation along the Livingstone-Ballantyne Cove Fault. This thrust is not a single fault but a zone consisting of three separate high-angle fault zones striking N. 85° E. to N. 85° W. and dipping about 70° to the north. The faults are best exposed near Ballantyne Cove on the east side of the Cape.

The northernmost fault plane consists of a hard mylonite zone 6 in. thick which is exposed on the shore north of the Cove. On either side of the mylonite, there is 4 in. of stretched pebbles that are coated by hematite. The middle fault is exposed both on the shore and inland in the stream bed of Cove Brook several hundred feet from the shore. The southernmost fault is well exposed in a cliff on the north side of the road at Ballantyne Cove. These fault planes are not as clearly exposed at Livingstone Cove. There, the only evidence for faulting is a hard black conglomerate with stretched pebbles. The black color is due to hematite that coats the pebbles.

Inland, the fault zone forms an east-west valley that extends between the two coves. The valley probably results from easier erosion of material in the fault zone. In both coves, wave erosion of the fault zone has produced the prominent indentations of the shoreline that comprise Livingstone and Ballantyne Coves (Pl. 12).

The conglomerate of the Horton Formation north of the fault has been folded as a result of the faulting into minor anticlines and synclines 30 to 100 ft across. This folding has affected the beds up to 400 ft north of the fault zone.

The relative motion along the fault is difficult to determine. Evidence of relative motion indicates a high-angle thrust fault, the north side overriding the south side with a predominantly vertical component of movement. This hypothesis is based on the following evidence: Horton Formation to the north is in fault contact with McAras Brook Formation, which is either Horton or Windsor in age; to the south, the small dragfolds in the Horton suggest an upward motion of the northern block, the elongation of stretched pebbles near the mylonite zone being parallel to the strike of the fault plane.

The actual amount of displacement is difficult to obtain. Kaminsky (1953) estimates a stratigraphic displacement of 5,000 ft. This figure is based on his estimates of the thickness of the conglomerate of the Horton Formation, 4,000 ft, and the Windsor Formation plus the McAras Brook Formation, 1,000 ft. The age of the faulting is post–McAras Brook.

Along the west side of Cape George (Pl. 12) from Malignant Cove to Livingstone Cove, the shoreline is straight, paralleling the northeasterly extension of the Hollow Fault which would be just offshore. North of Livingstone Cove, the shoreline protrudes to the west and is not so straight as it is south of the cove. This change in direction of the shoreline which occurs at Livingstone Cove is probably the result of the Livingstone–Ballantyne Cove Fault. Thus the Livingstone-Ballantyne Cove Fault would be younger than the Hollow Fault, and displaces the Hollow Fault at Livingstone Cove.

The conglomerate of the Horton Formation north of Livingstone and Ballantyne Cove is cut by numerous minor faults. The amount of displacement along these faults is difficult to determine, as the conglomerate has a relatively uniform appearance throughout. Minor faults are more numerous near Ballantyne Cove, School Brook Cove, and Cormorant Cliff Cove. Andesite dikes, which cut the conglomerate, are found along the northern shore of the Cape and are probably related to the flows in the McAras Brook Formation. At the northern end of the Cape are two large brecciated fault zones that strike east-west and have a near-vertical dip. One is found in School Brook Cove and the other in Cormorant Cliff Cove.

The structure and stratigraphy of these two fault zones is discussed in detail below. Relative motion along the faults could not be determined. The fault zones contain blocks (as much as 50 ft across) of older rocks that have been faulted into contact with the Horton Formation. The faulted blocks consist of Middle or Upper Ordovician quartzite, hornblende diorite, Silurian quartzites unlike those known elsewhere in Nova Scotia, Silurian Moydart Formation, Silurian-Devonian Stonehouse Formation, and Devonian Knoydart Formation. These faulted blocks form a mega-breccia from which a systematic structure and stratigraphy can be determined. The beds generally strike to the north or northeast and dip to the west. The general stratigraphic sequence is oldest (Ordovician) rocks to the east and younger rocks to the west.

The Ordovician quartzites are much more folded and faulted than are the Silurian and younger rocks. Unlike the younger rocks, the Ordovician rocks have been metamorphosed to quartzites, and have been intruded by hornblende diorite. Thus it is inferred that the Taconic orogeny resulted in deformation, metamorphism, and intrusion of the Ordovician strata before Silurian beds were deposited.

Within the fault zone in School Brook Cove, there is an angular unconformity separating the Lower Devonian Knoydart Formation below, attitude N. 28° E., 50° W., and the Mississippian Horton Formation above, attitude N. 25° E., 35° W. This angular relation is the result of the Acadian orogeny which occurred after the Lower Devonian strata were deposited.

Southeastern Area

The southeastern region is bounded on the northwest by the northeasterly striking fault that intersects the shore at Marsh Cove.

The rock types in this area consist of rhyolite, sandstone of the Horton Formation, and Windsor Formation (limestone, conglomerate, and sandstone).

The fault that bounds the area on the northwest is not exposed at any point. The only evidence for the fault is a prominent topographical scarp that strikes in a northeasterly direction and can be traced to within 400 ft of the shoreline. Northwest of the scarp, the land is 500 to 1,000 ft in elevation. This scarp marks the dividing line between Browns Mountain Group to the northwest and Mississippian strata to the southeast. At Marsh Cove, where the fault should intersect the shoreline, there are no outcrops for an interval of 260 ft. The first outcrop south of the covered interval is the pitted yellow (A_2) limestone of the Windsor Formation, which has been deformed into small drag folds. North of the covered interval are conglomerates and basalts of the McAras Brook Formation, and these rocks are fractured and sheared. The conglomerate in the first outcrop north of the interval contains pebbles with shiny black coatings of hematite. Float within the covered interval consists of blocks of gypsum and red sandy siltstone which were probably derived from the Windsor Formation.

North of the covered zone, outcrops of McAras Brook Formation along the shore exhibit a brecciated zone and a minor N. 60° E.-striking fault that are considered to be associated with the northeasterly striking scarp-forming fault.

The relative movement of the fault has not been determined. The only evidence present is that older rocks, such as Browns Mountain Group, are on the northwest side and younger rocks, Carboniferous, are on the southeast side.

Near the source of Wilkie Brook, there are a series of rhyolite flows that strike northeast and have a vertical dip. These rhyolites are on the southeast side of the fault.

Sandstone of the Horton Formation crops out in the lower part of Wilkie Brook and along the shore near the mouth of the brook. The sandstone is cut by minor faults and contains several minor folds. On the shoreline 1,500 ft north of Wilkie Brook, basal sandstone of the Windsor Formation rests with angular unconformity upon the Horton Formation. Here, the Horton Formation strikes N. 20° W., dips 37° E. and the Windsor Formation strikes N. 50° W., dips 28° N. The angular uncomformity between the Horton Formation and the Windsor Formation is also exposed halfway up Wilkie Brook, thus it is inferred that the Horton Formation was slightly folded before the deposition of Windsor sediments and that it may have had an undulating relief in the form of low hills ranging from tens of feet to hundreds of feet in height at the time the Windsor Formation was deposited.

Near the headwaters of Wilkie Brook, the relation between the Windsor Formation and the rhyolite is either a fault or an angular unconformity. Beds of the Windsor Formation strike parallel to the contact and dip 70° E. to vertical. If the contact is a fault, then it must be a branch of the northeast-striking scarp-forming fault to the northwest (described below). However, if the contact is an angular unconformity, then the observed relations can be explained by the presence within the Mississippian depositional basin of a topographic knob or hill of rhyolite that has been buried by onlap of first the Horton Formation and then the Windsor Formation.

Sedimentation during the Carboniferous

South of Livingstone Cove, basal conglomerate of the Horton Formation contains smaller clasts and its thickness is much less than is the case at the northern tip of Cape George. At School Brook Cove, the thickness of the basal conglomerate as well as the size of its boulders (up to 5 ft) indicate that the beds were deposited rapidly and that the sediment was derived from a nearby source area to the north or northwest (now covered by Northumberland Strait).

The thickness of the Horton conglomerate is extremely variable from locality to locality. The thickest section of the Horton Formation at Cape George is the southern limb of the syncline north of Livingstone Cove. The source rock for conglomerate of the Horton Formation is to be sought chiefly in the pre-Silurian metamorphic and igneous rocks. The conglomerates appear to be brought in from both the north or northwest and the south and deposited in a basin. The area south of Livingstone Cove consisting of Browns Mountain Group appears to have been an area of positive relief during Horton time and provided pebbles and matrix for the conglomerate. The conglomerates appear to be a fault-scarp talus deposit, derived from an area that rose quickly and was quickly eroded.

Boulders containing Middle or Upper Ordovician fossils have been found in this basal conglomerate of the Horton Formation at Cormorant Cliff. The only other known Ordovician strata reported from this part of Nova Scotia are the shelly Middle or Upper Ordovician fossil localities of School Brook Cove (F-7, 10, 11). The absence of similar lithologic rock types in the pre-Silurian Browns Mountain Group of Antigonish and Pictou Counties suggests that the bulk of the Middle or Upper Ordovician quartzites were originally located north or northwest of Hollow Fault, which bounds the northwestern coast of Cape George.

The probable fault scarp could have been produced by the same stress system that produced the movement along the fault zones at School Brook Cove and Cormorant Cliff Cove. Some movement could have taken place within the School Brook Cove and Cormorant Cliff Cove Fault zones, before the deposition of the Horton Formation. Considerable movement did take place after the deposition of the conglomerate, as the conglomerates are faulted.

As the basin was filled up, the relief of the source area decreased and the sediments which were deposited in the basin gradually became finer-grained. Near the end of Horton time, sandstones were deposited within the basin, and there may have been an island within the basin, made up of the rhyolite that now crops out at the head of Wilkie Brook. As the basin filled up, the sediments would have been deposited farther up the slopes of the island in a sedimentary onlap over the rhyolite. At the end of Horton time, the sediments within the basin were deformed slightly.

After the slight deformation of the Horton Formation, Windsor sandstone, conglomerate, and limestone were deposited in the basin. During the time of deposition of the Windsor Formation, the rhyolite at the head of Wilkie Brook might still have been exposed as an island. As a result of basin-filling and the sedimentary onlap of Windsor Formation, the island would have been covered to a greater extent than after deposition of the Horton Formation.

The Windsor Formation at Wilkie Brook contains more conglomerate and sandstone and less limestone than is found in other Windsor sections of Nova Scotia. In addition, there are mudcracks, and algal-reef limestone rather than the thin-bedded limestone more typical of the Windsor Formation. Large cross-beds as much as 8 ft long occur near the angular uncomformity with the Horton Formation. This evidence is interpreted to mean that these rocks were deposited in a nearshore or shoreline environment, rather than the marine basin environment more typical of the Windsor Formation.

Murray (1956) noted that the Windsor Formation at Wilkie Brook is different from the normal Windsor sections as suggested by the following quotation: "Subzone A of the Windsor Group undergoes a facies change from the normal limestone, gypsum and siltstone assemblage present throughout Nova Scotia, into red argillaceous algal and other typical shoreline deposits and red sharpstone conglomerate. The normal sequence loosely called central basin facies is easily mapped as a unit approximately equivalent to a shoreline facies, the loosely called marginal basin facies, is mapped as the Wilkie Brook Formation."

During the Mississippian, either during Horton or Windsor time, basaltic flows and conglomérates of the McAras Brook Formation were deposited in the basin.

The major faulting in the area occurred after Carboniferous sedimentation ended.

MIDDLE OR UPPER ORDOVICIAN

Hornfels (O_1 of Pls. 13, 14)

Lithology. The hornfels consists of interbedded sandstone layers 4 to 6 ft thick and shale layers 2 to 3 ft thick that have been baked by the intrusion of hornblende diorite.

The baked sandstone is now a clear white quartzite. Outlines of individual grains are indistinguishable with the naked eye, but the surface of the rock has a granular appearance. The rock is hard, dense, massive, and does not have any bedding features except for one bed that contains cylindrical tubes perpendicular to bedding. These tubes have resulted from the boring of wormlike animals into the sediment, and now are filled with well-cemented medium sand-sized quartz grains.

The shales have been baked to a hard dense fine-grained rock which is composed primarily of quartz. Bedding has survived the baking and remains as brown and black color bands. The beds are ¼ to 2 inches in thickness.

Thickness. The minimum thickness of the hornfels is 40 ft.

Lower Contact. Neither the actual thickness nor the lower contact is known because of the complex intrusive relation of the hornblende diorite, the shearing movements resulting from intrusion, and later Carboniferous folding and faulting. The hornfels is the lowest in the stratigraphic series found at the tip of Cape George.

Age. Middle or Upper Ordovician fossils are found in the hornfels at the mouth of School Brook. The fossils are found in the purplish-black fractured shale, locality (F-7) of the map (Pl. 13). The shale weathers readily and contains limonite which leaches out, leaving a crumbly porous rock. The fossils include poorly preserved specimens of the brachiopods *Rhynchotrema* and *Rafinesquina,* poorly preserved tentaculitids, and a fragmentary trilobite pygidium.

Quartzite (O_1 of Pls. 13, 14)

Lithology. The quartzite is indistinctly bedded, massive, sugary in appearance, and contains subangular, medium-sized quartz grains. In color, it is dirty-gray to white, and the grains are well sorted. The rock is well cemented by quartz and calcite and is hard when fresh, crumbly when weathered. A small percent of chloritic argillaceous material is present. Some parts of the quartzite are mottled gray and white due to a slight concentration of the argillaceous material, and the rock contains about 1 percent limonite. The quartzite beds are 5 in. to 1½ ft thick. The lower part of the quartzite contains thin beds, 2 in. thick of dark-green shaly layers, which consist of greenish silt and clay-sized material. Oscillation ripple marks occur on the surfaces of several of these beds. The planes of a prominent joint system are parallel to the bedding.

The quartzite has been baked by the intrusion of the hornblende diorite.

Thickness. The thickness of the quartzite is 30 ft.

Lower Contact. The lowermost exposed quartzite is intruded by hornblende diorite.

Age. The quartzite lies above the hornfels but according to the fossil evidence, it is the same age as the hornfels. Poorly preserved casts of Middle or Upper Ordovician strophomenid and rhynchonellid brachiopods are found in the quartzite at localities (F-10) and (F-11).

Incorporated within the Horton conglomerate of Cormorant Cliff is a large quartzite boulder from which Raymond Fletcher has collected fossils of Middle or Upper Ordovician age. The fossils include *Kaierina*? and an unidentified atrypacean brachiopod that has more primitive cardinalia than would be expected in rocks of Silurian age.

The occurrence of shelly Middle or Upper Ordovician fossils in Nova Scotia is of considerable interest because it indicates the presence of a shallow-water environment in this region. The absence of lithologically similar rock types in the pre-Silurian Browns Mountain Group of Antigonish and Pictou Counties suggests that the original location of the bulk of the Middle or Upper Ordovician quartzites may have been northwest of the Hollow Fault that bounds the northwestern coast of Nova Scotia in the Cape George region.

Dark-Colored Granitic Rock (Hornblende Diorite) (O_2 of Pls. 13, 14)

This unit is holocrystalline phaneritic, with grains to 2 cm in size. The texture of the rock is hypidiomorphic-granular. In general, the rock is dark brown to black, but a part of the intrusive is pinkish due to a large percentage of orthoclase with a corresponding lack of amphiboles. This granitic rock (hornblende diorite) consists of plagioclase (35 percent), orthoclase (30 percent), hornblende (35 percent), and pyrite. Ordovician strata are intruded by the unit, but it does not cut rocks stratigraphically above the Ordovician in this region.

SILURIAN
NON-ARISAIG SERIES

Quartzite (S_1 of Pls. 13, 14)

Lithology. The quartzite is white to grayish white on both weathered and unweathered surfaces. It is massive and well cemented by secondary quartz. The grains are medium sand-sized, well sorted, closely packed, and rounded. The rock contains layers of finer sand which are grayer in color and also contains a few limy quartzite beds, about 1 in. thick, that are dark gray-blue. Bedding is easily seen because of thin laminations, color variations due to changes in grain size, and cross-bedding.

The basal beds are grayish-white with irregularly shaped patches of clean white quartzite which occur only in certain stratigraphic horizons of 1½ to 3 ft thickness. These patches appear to have resulted from bottom-crawling animals or worms that destroyed the original stratification or laminations of the beds and mixed layers of grayish-white and clean-white sand into the irregularly shaped patches that would not have been produced by sedimentary processes.

Thickness. The exposed quartzite unit is 50 ft thick.

Lower Contact. The lower contact is not exposed, but the lowest beds exposed are faulted against the hornblende diorite.

Age. Fossils have not been found in this quartzite, but since it is directly below and conformable with the *Scolithus* (worm tube) quartzite, which contains Silurian fossils, it is believed to be of Silurian age.

Scolithus (Worm Tube) Quartzite (Fig. 22) (S_2 of Pls. 13, 14)

Lithology. The *Scolithus* (worm tube) quartzite consists of well-sorted, closely packed, well-rounded, medium sand-sized quartz grains. The quartzite is very hard and difficult to break. Some of the beds are grayish white and have mottled dark-gray areas containing about 5 percent silt. The quartzite is white on both fresh and weathered surfaces, but those beds that contain small percentages of silt weather to grayish brown and become friable.

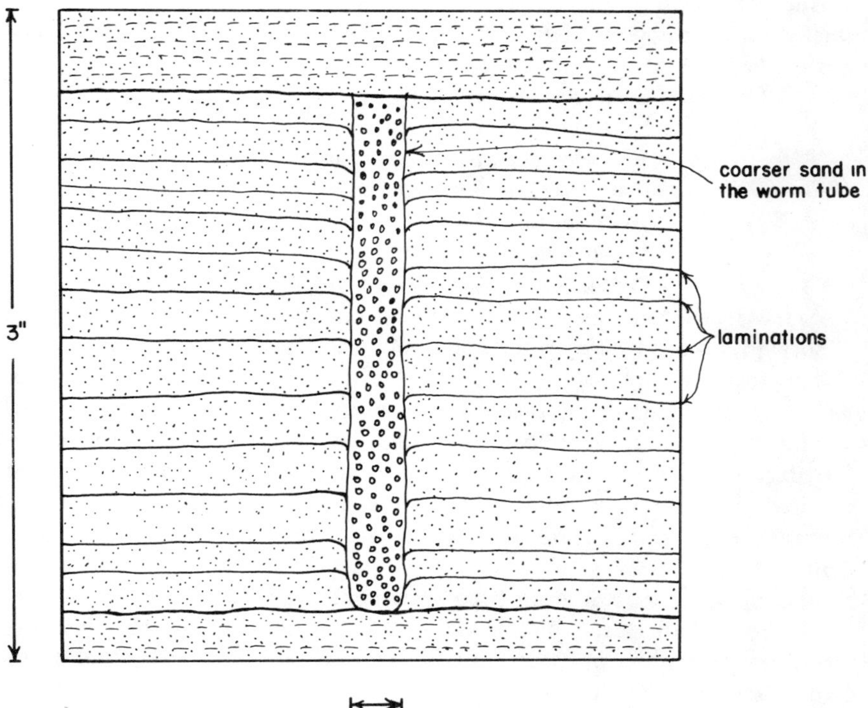

Figure 22. Sketch of worm tube in *Scolithus* (worm tube) quartzite.

Beds alternate between clean white quartzite, 2 to 3 ft thick, and gray silty quartzites, about 2 ft thick, containing *Scolithus*. The worm tubes tend to be perpendicular to bedding, and are 3 in. long and a ¼ inch in diameter. They are filled with rounded and well-sorted quartz grains that are of grain size coarser than that of the surrounding quartzite. In the quartzite, there are thin laminations caused by small changes in grain size. These laminations curve downward where they are cut by a worm tube, which provides a good indicator for distinguishing top and bottom of a bed (Fig. 22). At the bottom of the *Scolithus* quartzite, there is a 2-ft-thick fossiliferous bed of medium-grained, subangular, light-gray quartzite. This lithology is calcareous and contains well-rounded black pellets that are ¼ inch in diameter.

Thickness. The *Scolithus* (worm tube) quartzite is 50 ft thick.

Lower Contact. The worm-tube beds are separated from outcrops of the (S_1) quartzite by a covered interval 2 ft wide. The *Scolithus* quartzite lies above and is conformable with the (S_1) quartzite unit (of Pls. 13, 14).

Age. The fossils found at the bottom of the *Scolithus* quartzite consist chiefly of brachiopods, including dalmanellids and chonetids of Silurian aspect. The following fossils were identified:

(F-17) *Hyattidina northumberlandensis*
Salopina submedia
"Camarotoechia" cf. *squamifera*
Protochonetes novascoticus

(F-18) *Shaleria gilpeni*
Isorthis fornicatimcurvata
Protochonetes novascoticus

A few fish fragments were recovered from the *Scolithus* quartzites, but T. Ørvig reports (1959, written commun.) that they are not identifiable.

The fossil assemblage, particularly the dalmanellid, is not correlatable with any of the units in the lower half of the Arisaig Group although similar to those in the upper half. The "Red Band" at the top of the Moydart Formation is recognizable at School Brook Cove, although its detailed stratigraphy is different from that at Arisaig. The Knoydart and Stonehouse Formations are also present at School Brook and Cormorant Cliff Coves.

It is puzzling that Silurian lithologies similar to those of the McAdam, Ross Brook, and Beechhill Formations, which underlie the Moydart elsewhere in Antigonish and Pictou Counties, have not been recognized in the Cape George area. Even more puzzling is the presence of a sequence of Silurian rocks which are lithologically unlike anything else in the Silurian of Nova Scotia.

Red and Grayish-White, Cross-Bedded Sandstone (S_3 of Pls. 13, 14)

Lithology. The lower part of this unit consists of grayish-white, well-cemented, medium sand-sized sandstone. The grains are well-rounded, well-sorted, and are mostly closely packed quartz cemented by calcite.

Higher in the section, the quartz grains are much larger, and the rock is less calcareous and not so well cemented. Here, the sandstone is yellowish white and contains about 10 percent feldspar.

Sixty feet above the base of this unit, the yellowish-white sandstone abruptly changes to a dark-red, 15-ft-thick sequence of alternating sandstone and siltstone. Small amounts of hematite color the rock and provide some of the cementation. Beds are 4 in. to 1 ft in thickness. The red sandstone consists of subangular, medium sand-sized quartz grains. The grains are poorly sorted and the rock contains about 25 percent silt, plus about 2 percent muscovite. The muscovite flakes are evenly distributed throughout the sandstone parallel to bedding.

The uppermost part of the section consists of yellowish-white sandstone alternating with red interbedded sandstone and shale.

The whole section has well-developed laminations ¼ to ½ in. thick with well-developed cross-bedding.

Thickness. The exposed thickness of the unit is 130 ft.

Lower Contact. There is a 200 ft covered interval between the cross-bedded sandstone and the nearest *Scolithus* quartzite outcrop. Location and attitude of bedding of both units indicates that they are conformable and that the cross-bedded sandstone (S_3 of Pls. 13, 14) lies above the *Scolithus* quartzite.

Age. No fossils have been found in the cross-bedded sandstone, but it is believed to be the same age as the *Scolithus* (worm tube) quartzite (S_2 of Pls. 13, 14).

Gray-Blue Siltstone (S_4 of Pls. 13, 14)

Lithology. This siltstone is gray-blue on fresh surfaces and light greenish blue on weathered surfaces. The grains are poorly sorted, fine to coarse silt size. Folding and shearing have deformed most primary features, so that they are not recognizable. Interbedded with the siltstone are bands of limy sandstone, 6 to 8 in. thick. These beds are greenish gray on fresh surfaces and grayish white on weathered surfaces.

The rock consists of poorly sorted, subangular to subrounded grains of medium sand size. The grains consist of quartz, feldspar, rounded fragments of calcite (which are broken-up fossil fragments), subrounded grains of glauconite up to ⅓ cm in width, as well as brachiopods, ostracodes, cephalopods, and possible fish fragments. Scattered throughout the limy sandstone beds are patches of grayish-green siltstone, some of which are pyritic.

The rock is well cemented by secondary calcite and has little void space.

Thickness. The thickness of the unit is difficult to obtain because of faulting and complex folding. The minimum thickness is 15 ft.

Lower Contact. The unit is bounded on three sides by faults (Pl. 12). The locations of the outcrops indicate that it lies above the red and grayish-white cross-bedded sandstone (S_3 of Pls. 13, 14).

Age. The fossils found in the unit indicate a Stonehouse age. The unit is placed in the non-Arisaig Silurian (Table 12) as the uppermost unit.

The following fossils were identified from locality (F-19).

Brachiopods
Delthyris rugaecosta
Isorthis fornicatimcurvata
rhynchonellid
Ostracodes
Aparchites sinuatus

The Silurian quartzite, the *Scolithus* quartzite, the red and grayish-white cross-bedded sandstone, and the gray-blue siltstone of School Brook Cove are lithologically unlike any other known Silurian of Nova Scotia. The fauna of the *Scolithus* quartzite is not correlatable with any unit in the Arisaig Silurian or with any known Silurian of Nova Scotia. The occurrence of the *Scolithus* quartzites and the other three units at Cape George and their absence elsewhere can best be explained by inferring their presence to the northwest on the other side of the Hollow Fault.

The following are three possible correlations of *Scolithus* quartzites with the Silurian of Arisaig:

1. The unit corresponds to a facies change (nearshore or shoreline deposit) of a part of the Arisaig lower Silurian, either the Beechhill Cove Formation, the Ross Brook Formation, or the McAdam Brook Formation.
2. The unit is pre-Beechhill Cove Formation and its sediments were deposited at Cape George before deposition began at Arisaig during the Silurian.
3. The *Scolithus* quartzite was deposited at Cape George during a short time in the Silurian when no deposition was taking place at Arisaig.

SILURIAN AND DEVONIAN ARISAIG GROUP, STONEHOUSE-MOYDART

Portions of the mega-breccia at School Brook Cove and Cormorant Cliff Cove have been assigned to the Stonehouse and Moydart Formations

TABLE 12. Stratigraphy of School Brook and Cormorant Cliff Coves

	Estimated Thickness (ft)
Carboniferous	
Horton conglomerate (Cc)	
Devonian	
Knoydart	
Mottled red and green siltstone (Dk_2)	min. 20
Sheared red sandstone and shale (Dk_1)	——
Silurian and Devonian	
Arisaig Series	
Stonehouse	
Gray-blue siltstone (Ds_2)	min. 25
Silty sandstone (Ds_1)	min. 60
Silurian	
Moydart	
Red Band	
Nodular sandy limestone (Sm_6)	8 to 12
Red shaly sandstone (Sm_5)	½
Red mottled sandstone (Sm_5)	2
Nodular red sandstone (Sm_5)	1½
Red sandstone (Sm_5)	8
Pre-"Red Band" Moydart	
Massive sandstone (Sm_4)	5
Interbedded siltstone and quartzite (Sm_4)	4
Siltstone and mudstone (Sm_4)	8
Quartzite (Sm_4)	½ to 1
Blue siltstone (Sm_3)	5
Black and red sandstone (Sm_2)	2 to 3
Fine-grained sandstone (Sm_1)	3 to 6
Blue siltstone (Sm_1)	min. 70
Non-Arisaig Series	
Gray-blue siltstone (S_4)	min. 15
Red and grayish-white, cross-bedded sandstone (S_3)	min. 130
Scolithus (worm tube) quartzite (S_2)	50
Quartzite (S_1)	min. 50
Dark-colored granitic rock (O_2) (hornblende diorite)	
Middle or Upper Ordovician	
Quartzite (O_1)	min. 30
Hornfels (O_1)	min. 40

because of the lithologic similarities between the red beds at School Brook Cove and the "Red Band" at the top of the type Moydart Formation near Arisaig (Fig. 30). There is not a bed-to-bed correlation between the "Red Band" of the Arisaig section and the "Red Band" exposed at Cape George. The basis for correlation is the presence in both areas of red sandstone, algal nodules, and red shaly sandstone. Correlation is further strengthened by a comparison of thicknesses in the two areas: 20 to 24 ft at Cape George and 30 ft at Arisaig (Fig. 30).

The "Red Band" at the top of the Moydart Formation has also been recognized in an area in Pictou County (Fig. 30). Similarly, its bed-for-bed detailed stratigraphy is not the same as that of the "Red Band" at Arisaig, but the presence of algal nodules and algal columns in red siltstone and shale is also considered to strongly support the correlation. Algal nodules and columns become more abundant as one goes eastward from Pictou County into the Arisaig area (R. H. Maehl, 1959, oral commun.). Farther to the east, at Cape George, they are more abundant than at Arisaig. The top of the "Red Band" of the Arisaig section is the upper contact of the Moydart Formation with the Stonehouse Formation; therefore the top of the "Red Band" at Cape George is taken to be the upper contact of the Moydart Formation there.

Delthyris rugaecosta, which occurs in the Stonehouse Formation of Arisaig and of Pictou County, is also found at Cape George above the "Red Band." The shales in which the fossil occurs can also be correlated with the Stonehouse Formation because of their lithology and stratigraphic position.

Ami (1895) has described twenty species collected by Hugh Fletcher in 1886 from the tip of Cape George. The collections were made in the Silurian and Devonian outcrops of School Brook and Cormorant Cliff Coves. The fossil localities from the Fletcher report could not be related with the fossil localities plotted on the detailed maps of School Brook and Cormorant Cliff Coves.

The fossils listed by Ami are:

- Annelids
  - *Serpulites longissimus*
  - *Tentaculites niagarensis*
  - *Tentaculites canadensis*
- Brachiopods
  - *Discina novascotia*
  - *Discina fletcheri*
  - *Discina orientalis*
  - *Lingula rectilatera*

Lingula sp. resembles *Lingula perovata*
Rhynchonella formosa
Lamellibranchiata
Orthonota equilatera
Modiolopis exilis
Nuculites (Clidophorus) erectus
Nuculites sp.
Gastropoda
Bucania sp.
Holopea reversa
Cephalopoda
Orthoceras sp. cf *annulatum*
Orthoceras sp.
Ostracoda
Leperditia sp.
Vertebrata
Pisces
Onchus (?) sp.

Discina novascotia, Tentaculites canadenis, Discina fletcheri, and *Discina orientalis* were described as new species by Ami (1895):

In age they correspond most probably with those of Divisions "C" [Stonehouse Formation] and "D" [Moydart Formation] of the Arisaig Section of Nova Scotia and to the Niagara and Lower Helderberg of New York State and Ontario. They are therefore referable to the Silurian epoch. Some of the forms examined appear to be closely related to species from various localities in the State of New York, whilst others seem to be more nearly related to well-known European forms of Wenlock and Ludlow age. The presence, in tolerable abundance, of shells of the genus *Discina* is in itself quite characteristic, and a fact to be noted. Besides the twenty forms described or enumerated in this paper, there are numerous fragments and obscure remains of fossils which although too imperfect for specific identification, nevertheless may serve to point out the occurrence and association of *Chonetes, Rhynchonella, Retzia, Murchisonia, Homalonotus*?. The presence of a number of Ludlow and Lower Helderberg fossils tend to indicate a horizon well up in the Silurian System.

Ami (1895) also noted: "These fossils from Cape George seem to indicate a zone or horizon which differs considerably from those of the Arisaig Coast collected by Mr. Weston in 1886, and which have also been examined. This is probably due to local causes."

This difference between the Arisaig and the Cape George specimens may be due to the fact that some of the Cape George specimens were collected from the Silurian *Scolithus* quartzites.

Kaminsky (1953) collected and Boucot identified the following fossils from School Brook and Cormorant Coves. Kaminsky's localities were not precisely located on his maps.

Chonetes novascotius
Orthonota sp.
Modiolopsis
large pelecypod similar to *Pteronitella, Cuneamya arata*
brachiopod resembling *Pectorhynchia obtusiplicata*
Discina
Lingula
Tentaculites
several genera of small pelecypods
fish scales

MOYDART FORMATION (PRE-"RED BAND" MOYDART)

Blue Siltstone (Sm_1 of Pls. 13, 14)

Lithology. The lowest exposed beds of Moydart Formation consist of a grayish-blue, coarse-grained, poorly sorted siltstone that grades into a fine-grained sandstone. The rock is massive, calcareous, and poorly bedded. Beds are greatly fractured and some are fissile. Parallel parting planes and faint laminations indicate bedding orientation. The rock weathers to a light-blue color with rusty stains, and becomes friable.

Thickness. The amount of exposed blue siltstone is 70 ft.

Lower Contact. The lower contact of the Moydart Formation is not exposed.

Age. No fossils were found in the siltstone, but its position below the "Red Band" places it in the Moydart Formation.

Fine-Grained Sandstone (Sm_1 of Pls. 13, 14)

Lithology. The fine sandstone is massive, and contains well-sorted subangular quartz grains which are partially cemented by calcite. The rock is light gray and well cemented, with no visible void spaces on fresh surfaces. Weathered surfaces are light tan and are stained brown by limonite. In some areas, this stain is black. The weathered surfaces also have voids and pits resulting from the removal of calcite that cemented the grains. Throughout the sandstone, crinoid columnals readily weather out. In the upper third of the sandstone, there is a 4-in. bed of medium sand-sized quartz grains. This bed is massive, better cemented, harder, shows no bedding features, and weathers white.

Thickness. The fine sandstone unit is 3 to 6 ft thick.

Lower Contact. The lower contact is gradational with the blue siltstone.

Age. Crinoid columnals are the only fossils found. The sandstone is placed in the Moydart Formation because of the position of the unit below the "Red Band."

Black and Red Sandstone (Sm_2 of Pls. 13, 14)

Lithology. The black sandstone is coarse grained, poorly sorted, and loosely packed. Grains are subrounded to rounded and consist of quartz. The rock contains a few rounded black pellets ¼ inch in diameter. The rock is cemented by black magnetite and a small amount of hematite which fills the void spaces. The beds are massive, blocky, and they part on the bedding planes. Within the black sandstone, there is a bed of red hematitic sandstone of variable thickness (1½ to 6 in.) which grades and lenses into the black sandstone. The grains of the hematitic sandstone are similar to the grains of the black sandstone but are more tightly packed. The red sandstone contains blue phosphatic fish fragments and brachiopod casts which on weathered surfaces are marked by yellow-brown limonite. Quartz grains of the hematitic sandstone are coated with shells of hematite, and hematite also forms the cement, which contains 1 percent calcite. The red sandstone appears to result from oxidation of the black sandstone.

Thickness. The thickness of the black and red sandstone is 2 to 3 ft.

Lower Contact. The lower contact grades into the fine sandstone that underlies it.

Age. A large rhynchonellid has been identified from locality (F-3). The unit is placed within the Moydart Formation because of its position below the "Red Band."

Blue Siltstone (Sm_3 of Pls. 13, 14)

Lithology. The siltstone is blue on fresh surfaces and greenish blue on weathered surfaces. Grain sizes range from coarse silt to fine sand, and beds are ⅛ to ¼ in. thick. The rock contains 5 percent of finely divided muscovite flakes oriented parallel to the bedding planes. Cleavage is slaty and is parallel to the bedding plane.

Thickness. The unit is 5 ft thick.

Lower Contact. The lower contact of the siltstone grades into the black and red sandstone (Sm_2).

Age. No fossils were found. The unit is placed within the Moydart Formation because of its position below the "Red Band."

Quartzite (Sm_4 of Pls. 13, 14)

Lithology. The quartzite contains fine sand-sized grains, well sorted, subangular, and closely packed. It is very well cemented by secondary quartz and has very little void space. The quartzite is blue on fresh surfaces and light tan on weathered surfaces. Crinoid columns are scattered throughout the bed. The top surface of the bed contains many black, rounded pellets, ¼ inch in diameter, and a few blue phosphatic fish scales. The top surface weathers to a deep tan with the pellets protruding above the surface.

Thickness. The bed is ½ to 1 inch in thickness.

Lower Contact. The lower contact is sharp and is conformable with the blue siltstone (Sm_3).

Age. The unit is placed within the Moydart Formation because of its position below the "Red Band" and because black phosphatic fish fragments and scales have been found at locality (F-5).

Siltstone and Mudstone (Sm_4 of Pls. 13, 14)

Lithology. The rocks grade from dull-red coarse siltstone to dull-red fine mudstone and consist of quartz (95 percent) and muscovite (5 percent). In general, bedding and cleavage are poorly developed, but at the base of the unit, a 2-ft gray-green section of harder, coarse-grained siltstone exhibits good bedding. These siltstone beds are ½ to 1½ in. thick.

Thickness. The unit is 8 ft thick.

Lower Contact. The unit is presumed to lie conformably on top of the 1-in.-thick quartzite bed (Sm_4). The precise stratigraphic position of the unit is in question, as it is cut at the top and bottom by faults.

Age. Because the strata are presumed to be above the quartzite (Sm_4) bed, the unit is considered to be of Moydart age.

Interbedded Siltstone and Quartzite (Sm_4 of Pls. 13, 14)

Lithology. Gray-blue, coarse siltstone beds ½ to 1 in. thick are interbedded with dark gray-blue quartzites that are 1 to 2 in. thick. The quartzites are fine-grained, hard, well cemented, and well sorted. The interbedding gives the rock a flaggy appearance. Top surfaces of the siltstone beds have markings that may have been made by worms.

Thickness. The unit of interbedded siltstone and quartzite is 4 ft thick.

Lower Contact. The bottom of the unit is cut by a fault, but it is presumed to lie conformably above the siltstone and mudstone.

Age. This unit is part of the Moydart Formation.

Massive Sandstone (Sm_4 of Pls. 13, 14)

Lithology. The sandstone is massive, hard, contains little void space, and is quartzitic. The grains are medium sand-sized, well sorted, subangular, and closely packed. The rock is light gray on fresh surfaces and lighter gray on weathered surfaces. Weathered surfaces crumble and powder easily. The upper section of the sandstone is light green and contains black pellets. In one outcrop (F-5), the sandstone contains a few brachiopods, crinoid columnals, and many blue phosphatic fish fragments.

Thickness. The sandstone is 5 ft thick.

Lower Contact. The sandstone lies conformably on the interbedded siltstone and quartzite (Sm_4) unit.

Age. The sandstone occurs in the stratigraphic section below the "Red Band" at School Brook Cove. Possible fish spines and linguloids have been found at locality (F-4), and outcrop (F-13) in Cormorant Cliff Cove has yielded a large collection of fish plates. Outcrop (F-13) is correlated with the massive sandstone found at School Brook Cove on the basis of lithologic similarity. Tor Ørvig of the Swedish Natural History Museum, Stockholm, has kindly examined some of the specimens from Cormorant Cliff Cove (F-13), and he reports as follows: "seems to consist of Thelodontid scales (of the *Thelodus parvidens* kind) and acanthodian spines and scales, the latter being referable to *Nostolepis* sp., and *"Gomphodus"* sp. The fauna, then, seems to be an Upper Silurian (Upper Ludlovian) one." This fish fauna is important because Upper Silurian fish are rare in the Appalachians, and also because it assists with the dating of the upper part of the Moydart Formation.

RED BAND OF THE MOYDART FORMATION (Fig. 23)

Red Sandstone (Sm_5 of Pls. 13, 14)

Lithology. The red sandstone has medium sand-sized grains, is loosely packed, and contains 2 to 3 percent muscovite. The grains are well rounded and poorly sorted. The rock weathers to a light red and is friable. The red color comes from hematite in the cement. This sandstone grades to a light green toward the lower part of the unit. The lower section of the sandstone contains black, rounded pellets up to ¼ inch in diameter, and a trace of secondary pyrite cubes which are ⅛ in. across. Well-developed joints are present.

Thickness. The red sandstone unit is 8 ft thick.

Lower Contact. The lower contact grades from massive sandstone (Sm_4) to red sandstone (Sm_5).

Age. The unit is part of the Moydart Formation "Red Band."

Nodular Red Sandstone (Sm_5 of Pls. 13, 14)

Lithology. The rock consists of a red, fine-grained sandstone with cal-

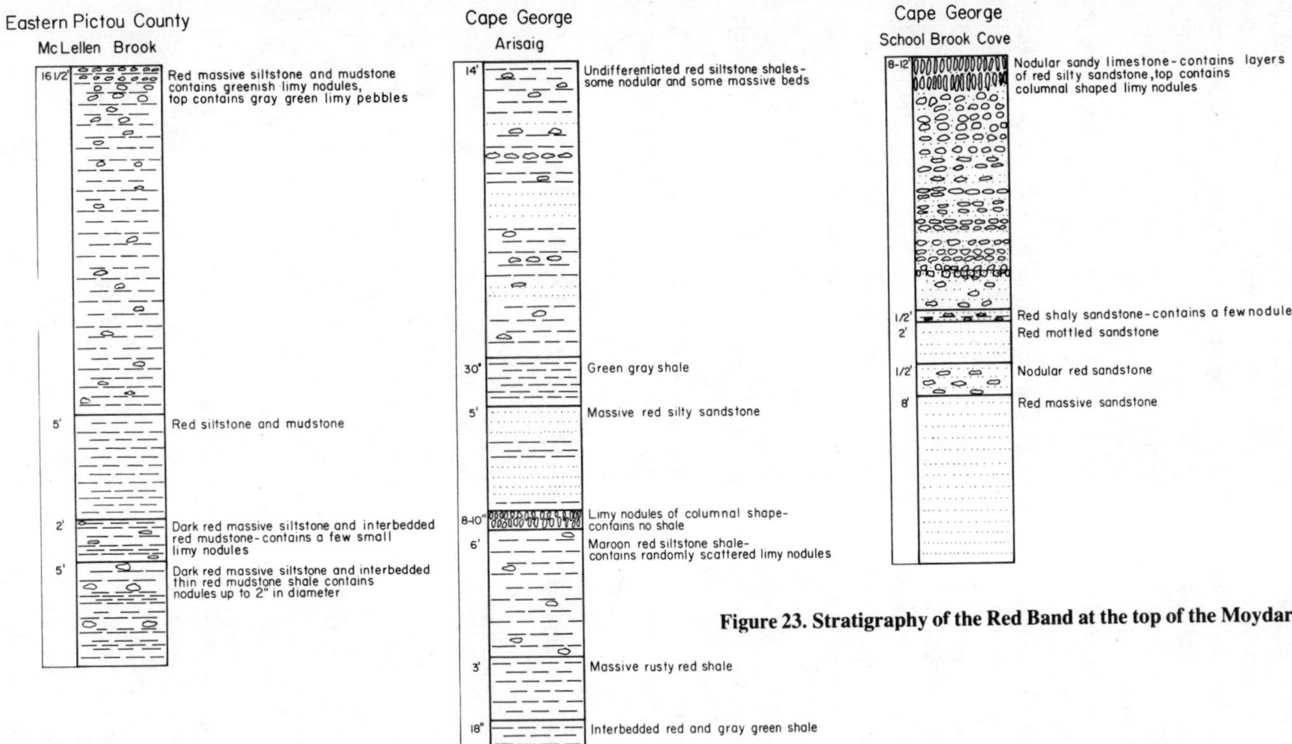

Figure 23. Stratigraphy of the Red Band at the top of the Moydart.

careous sandstone nodules in it. The nodules are as much as 3 inches in diameter and are harder and better cemented than the surrounding sandstone. The nodules consist of closely packed, fine sand-sized to coarse silt-sized quartz grains, plus 5 percent muscovite. They are cemented by calcite. On weathered surfaces, the nodules are yellowish green and knobby in appearance.

As the sandstone weathers, the nodules are left protruding above the surface. The red sandstone is coarser, less well sorted, and poorly cemented. The red color is produced by a small amount of hematite in the cementing material. A small amount of calcite also helps to cement the sandstone.

Thickness. The sandstone is 1½ ft thick.

Lower Contact. The lower part of the nodular sandstone (Sm_5) grades into the underlying red sandstone (Sm_5).

Age. The nodular red sandstone is part of the Moydart Formation "Red Band."

Red Mottled Sandstone (Sm_5 of Pls. 13, 14)

Lithology. This sandstone is medium sand-sized and contains quartz and a small percentage of muscovite (~3 percent). The rock is well cemented by calcite and by a small amount of hematite which produces the red color. The grains are poorly sorted, loosely packed, and well rounded. Many of the grains have a frosted appearance. Fresh surfaces of the sandstone are dull red, with calcite veins throughout. Weathered surfaces are mottled red and light green, with the light green being the more prominent color. The sandstone bed is massive with fine laminations.

Thickness. The bed is 2 ft thick.

Lower Contact. The mottled sandstone (Sm_5) grades into the nodular red sandstone (Sm_5) below it.

Age. The unit is part of the Moydart Formation "Red Band."

Red Shaly Sandstone (Sm_5 of Pls. 13, 14)

Lithology. The sandstone is red and contains quartz (80 percent), argillaceous material (15 percent), muscovite (5 percent), and a small amount of hematite. The grains are medium sand-sized, poorly sorted, and poorly cemented. The sandstone is cemented by hematite and argillaceous material. Throughout the bed there are lenses of harder and well-cemented sandstone. The sandstone weathers easily into small flat diamond-shaped fragments that are friable.

Thickness. The bed is ½ ft thick.

Lower Contact. The shaly sandstone (Sm_5) grades into the mottled red sandstone (Sm_5) below it.

Age. The red shaly sandstone is part of the Moydart Formation "Red Band."

Nodular Sandy Limestone (Sm_6 of Pls. 13, 14)

Lithology. The limestone consists of 40 percent sand grains and 60 percent calcite. The sand grains are of medium sand-sized quartz, well sorted, rounded, and loosely packed. The limestone is light gray, and in some areas it is mottled red and green. Bedding is poor. There are interlacing veins of white recrystallized calcite throughout which tend to be parallel to bedding. There are scattered layers of red silty sandstone within the limestone.

Nodules throughout the limestone account for 15 percent of the volume of the rock. The nodules are ½ to 3 inches in diameter, rounded, and harder than the surrounding limestone. The nodules are mottled red and green with bands of gray limestone and are cut by interlacing veins. Because they do not weather as readily as the surrounding rock, they protrude from weathered surfaces. The upper portion of this unit is redder and more shaly. The top has a columnar appearance due to tubes of quartz ¼ inch in diameter, normal to the bedding. (The columnar structures appear to have been built by algae. The nodules may also have been formed by algae.) The nodules are absent from 2 ft above the bottom of the limestone unit.

Thickness. The unit is 8 to 12 ft thick.

Lower Contact. The nodular limestone (Sm_6) grades into the red, shaly sandstone (Sm_5) below it.

Age. The limestone forms the top of the Moydart Formation "Red Band."

STONEHOUSE FORMATION

Silty Sandstone (Ds_1 of Pls. 13, 14)

Lithology. The silty sandstone is grayish green on weathered surfaces and bluish gray to green on fresh surfaces. The grains are medium sand-sized, rounded, and poorly sorted. The rock consists of 80 percent sand-sized quartz grains and 20 percent silt-sized grains. The upper part of the unit is poorly bedded. Toward the bottom of the unit, the beds are 4 in. to 1 ft thick, and parting parallel to bedding becomes pronounced. The lower beds are coarser grained and contain hematite, which produces a mottled dull-red and green color.

Thickness. The minimum thickness of the unit is 60 ft.

Lower Contact. The lower contact is a "knife edge" conformable contact with the Moydart Formation nodular sandy limestone (Sm_6). The upper part of the unit is cut by a fault.

Age. The age of the silty sandstone is assumed to be Stonehouse, as it is above the "Red Band." The "Red Band" of School Brook Cove correlates with the "Red Band" that is at the top of the Moydart Formation in the Arisaig section. No fossils have been found in the silty sandstone.

Gray-Blue Siltstone (Ds_2 of Pls. 13, 14)

Lithology. The siltstone is light greenish blue on weathered surfaces and gray-blue on fresh surfaces. The grains are poorly sorted, and grain size ranges from coarse silt to fine sand. The grains are well cemented by the clay particles which make up 5 percent of the rock, and by a small percentage of calcite. The siltstone contains limestone beds 2 to 4 in. thick. These limestone beds consist of 35 percent quartz sand grains, fossils, and fossil fragments. The fossils are brachiopods and ostracodes. Several individual outcrops of this unit were seen in both School Brook Cove and Cormorant Cliff Cove, but their relation to each other and to the underlying and overlying rocks is unknown due to faulting.

Thickness. The thickest unfaulted stratigraphic section of this unit is at least 25 ft thick.

Lower Contact. The siltstone appears to overlie the silty sandstone (Ds_1) of the Stonehouse Formation and to underlie the red sandstone and shale (Dk_1) of the Knoydart Formation. Because the unit is cut by so many faults, no contacts are seen, but the lower contact is assumed to be conformable with the silty sandstone (Ds_1).

Age. The fossils in this siltstone correspond to the fossils found in the Stonehouse Formation at Arisaig. The presence at Cape George of *Delthyris rugaecosta* places the unit in the Stonehouse Formation, because in Pictou County and the Arisaig area, *Delthyris rugaecosta* has been observed only in that formation.

The following fossils were recognized from Stonehouse localities in Cormorant Cliff Cove:

(F-1)

Brachiopods: *Isorthis fornicatimcurvata; Protochonetes novascoticus.*

Mollusca: orthoceroids; *Goniophora.*

Ostracodes: *Aparchites sinuatus; Kloedenia wilckensiana; Beyrichia* sp. indet.

(F-2)

Brachiopods: *Protochonetes novascoticus; Shaleria gilpeni; Isorthis fornicatimcurvata; Salopina* sp.

Ostracodes: *Aparchites sinuatus; Kloedenia wilckensiana; Beyrichia maccoyiana* cf. *sulcata*.

(F-8)

Brachiopods: *Protochonetes novascoticus.*

Ostracodes: *Aparchites sinuatus.*

(F-9)

Brachiopods: *Protochonetes novascoticus; Shaleria gilpeni; Delthyris rugaecosta.*

Arthropoda: trilobite (unidentified)

Mollusca: orthoceroid

Ostracodes: *Aparçhites sinuatus; Kloedenia wilckensiana; Beyrichia maccoyiana* cf. *sulcata; Beyrichia* sp. indet.

(F-12)

Brachiopods: *Hyattidina northumberlandensis; Sphaerirhynchia saffordi; "Camarotoechia"* cf. *squamifera; "C." rugosa; Protochonetes novascoticus.*

(F-14)

Brachiopods: *Protochonetes novascoticus.*

(F-15)

Brachiopods: *"Camarotoechia" rugosa.*

The ostracodes were identified by Murray Copeland; the brachiopods and other fossils were examined by Boucot. Copeland reports as follows (1959, written commun.): "All of these species have been obtained from strata of the Stonehouse Formation at Arisaig, Antigonish County, Nova Scotia. The absence of the distinctive *Beyrichia pustulosa* Hall, which is prevalent in the upper 400-500 feet of the type Stonehouse section may lead to the supposition that the Cape George beds are equivalent to some part of the lower strata of the Stonehouse Formation at Arisaig."

DEVONIAN
KNOYDART FORMATION

Sheared Red Sandstone and Shale (Dk_1 of Pls. 13, 14)

Lithology. The sandstone is light red on both fresh and weathered surfaces. It consists of rounded, well-sorted, medium sand-sized quartz grains, cemented by secondary quartz and hematite. The rock is colored by a small percentage of hematite. The sandstone has no bedding features and it is impossible to obtain a bedding attitude.

The shale is darker red than the sandstone, and it consists of fine silt and argillaceous material. The red color is due to the presence of a small percentage of hematite. The shale exhibits no bedding, only cleavage. Any bedding structures that were originally present in the shale have been eradicated by faulting and shearing. The shale is not uniform in color, but has patches and lenses of different shades of red due to variations in grain size. As grain size decreases, the shale becomes darker red. The lenses and patches are due to faulting which has produced slippage along the cleavage planes of the shale. Before faulting, the shale consisted of interbedded layers of coarser and finer grain sizes. Shearing and slippage along the cleavage plane destroyed the original bedding and produced the irregularly shaped lenses and patches.

The red sandstone occurs as lenses 1 to 3 ft long within the red shale. The unit consists of 85 percent red shale and 15 percent red sandstone.

Thickness. The thickness of the unit cannot be determined because the attitude of bedding has been destroyed by faulting and shearing.

Lower Contact. No contacts with other units, except fault contacts, have been found.

Age. The unit is assumed to be above the Stonehouse Formation. Since, in the Arisaig section, the Stonehouse Formation does not contain red beds and the Knoydart Formation does, the sheared red sandstone and shale at Cape George is placed in the Knoydart Formation because of its light red color and close association with the gray-blue siltstone (Ds_2) of the Stonehouse Formation. No fossils have been found in the unit.

Mottled Red and Green Siltstone (Dk_2 of Pls. 13, 14)

Lithology. The siltstone is poorly sorted and consists of coarse silt-sized grains of quartz (70 percent), plus fine sand-sized quartz grains (20 percent), and argillaceous material (10 percent). This unit contains red, green, and mottled red and green beds. The green beds are generally coarser grained than are the red beds. Bedding features are poorly developed within the shale, but the rock has parting that is parallel to bedding. Beds are 4 in. to 2 ft thick.

Within the unit are beds of fine-grained flaggy sandstone and fissile claystone. The top 3 ft of the unit is a light-green, fine-grained, silty sandstone.

Thickness. The minimum thickness of the unit is 20 ft.

Lower Contact. The shales are cut at the bottom by a fault and are presumed to be the highest part of the Knoydart Formation present in this area.

Age. The unit is placed in the Knoydart Formation because of its red color, and because of its close association with the Stonehouse Formation. It occurs below the angular unconformity with the Horton Formation. No fossils have been found in the siltstone.

STRUCTURE OF SCHOOL BROOK AND CORMORANT CLIFF COVES

Intrusion of Hornblende Diorite

Hornblende diorite intruded the sandstones and shales of Middle or Upper Ordovician age, and through heat and pressure baked the rocks into hornfels and quartzite. The hornblende diorite forced its way between the beds to form sills; it also forced its way into joints, cracks, and faults and widened them. The intrusion produced many xenoliths, some as large as 8 ft x 4 ft x 1 ft. Some of the xenoliths have been rotated. In places, the hornblende diorite contains color lineations of remnant sedimentary beds that may have resulted from lit-par-lit intrusion. Small offshoot stringers of hornblende diorite cut across the bedding.

At the outer periphery of the intrusion, there is a complex mixture of hornblende diorite, quartzite, and hornfels, which is folded and cut by numerous faults. Tracing the beds and reconstructing relative movement is difficult in this area of complex faulting. The folding and faulting is pre-Silurian, but later movement has taken place along large east-west faults which cut the hornblende diorite, quartzite, and hornfels.

Angular Unconformity

There is an angular unconformity between the Horton Formation of Mississippian age and the Knoydart Formation of Lower Devonian age. The attitude of the Horton Formation beds at the contact is N. 25° E., 35° W., and of the Knoydart beds N. 28° E., 50° W. The unconformity is assumed to be the result of deformation during the Acadian orogeny followed by later deposition of the conglomerate.

Fault Complex, Tip of Cape George

The northern tip of Cape George is a region cut by major east-west faults. Both School Brook Cove and Cormorant Cliff Cove are within this structurally disturbed region. The structure at each cove consists of a complex fault zone bounded on the north and south by major east-west faults that probably have both horizontal and vertical components of movement. Information regarding the location, relative movements, and amount of movement of many of the minor faults within these fault complexes can be obtained because of the several distinctive rock types and the preserved portions of stratigraphic column found in the area.

Each of the fault complexes can be considered as a mega-breccia containing blocks up to 30 ft wide and 50 ft long. The complex exposed at School Brook Cove is at least 300 ft wide and that at Cormorant Cliff Cove is 250 ft wide. Between these two complexes there are continuous outcrops of Horton Formation. The Horton Formation was not mapped in detail, but the conglomerate contains numerous minor faults and joints along which movement has taken place. Since the conglomerate is uniform and thick, the amount of relative direction of movement along faults is difficult to obtain.

Southern Boundary Fault at School Brook Cove

The mega-breccia exposed at School Brook Cove is bounded on the south by a major N. 88° W., 65° S. fault, along which most of the movement at the cove has occurred. This fault is the most prominent in the area, and it shows the greatest stratigraphic displacement of all the faults at the tip of Cape George. The fault separates Horton Formation on the south side of the fault from Ordovician hornfels and hornblende diorite on the north side. The fault zone is 5 ft wide. The southern 1½-ft thickness of the sheared fault zone consists of stretched pebbles of the Horton Formation. The strike of the long axes of these pebbles is horizontal and parallel to the strike of the fault zone. The northern portion of the sheared zone consists of a breccia of quartzite and hornblende diorite blocks derived from the north side of the fault. These blocks range from 3 in. to 4 ft long and 6 in. wide and have also been stretched and distorted. The center of the fault zone contains fragments of conglomerate and igneous rock in a matrix of silt-sized and sand-sized material together with rock flour. The stretched pebbles within the fault zone indicate displacement in the horizontal direction, but the age relations of rocks on either side of the fault suggest a vertical component of displacement. Relative movement in the horizontal plane was not determined, but the north side of the block moved up relative to the south side.

East of School Brook, this fault consists of three individual parallel traces in a zone 15 ft wide. A traverse across this zone from south to north reveals the following relations:

First fault—Horton Formation, south side; Ordovician quartzite, north side.

Second fault—Ordovician quartzite, south side; Horton Formation, north side.

Third fault—Horton Formation, south side; diorite, north side.

The Ordovician quartzite between the first and second faults and the Horton Formation between the second and third faults are fault slices within the southern boundary fault.

Northern Boundary Fault, School Brook Cove

The shoreline of School Brook Cove is controlled by another fault striking nearly west. In places, this is not one fault but a series of parallel faults. Silurian rocks are to the north and Ordovician rocks to the south of the fault. At the western end of School Brook Cove, the fault strikes east-northeast toward the sea. Between the parallel faults at the western end of School Brook Cove (Pl. 13B) are blocks of Silurian *Scolithus* (worm tube) quartzite (S_2), Silurian cross-bedded sandstone (S_3), and hornblende diorite (O_2). Surrounding some of these blocks is sheared Knoydart red siltstone (Dk_1) in which all bedding features have been destroyed. The red siltstone is within the fault zone and has been brought into its present position by movement of the faulted blocks. The relation and position of the different rock types as well as the orientation of minor faults suggest that the north block moved west relative to the south block. However, if this was the actual movement of the fault, the position of the Knoydart red siltstone would be difficult to explain since it is found faulted between Moydart Formation and *Scolithus* quartzite.

Mega-Fault Breccia at School Brook Cove

The beach lies between the two major fault zones of School Brook Cove. Outcrops are well exposed in this beach section and are cut by numerous normal, thrust, and strike-slip faults. These faults strike and dip in many different directions, and have cut the section into a zone of many fault-bounded blocks. This zone is termed a mega-breccia. Many of the blocks have been rotated and some blocks contain beds that have been drag-folded and then faulted. At the western end of School Brook Cove, a large drag fold, 70 ft from crest to trough, is cut by two faults striking N. 13° W. and N. 39° W. The relative movement of these faults was dextral, with the eastern side being displaced to the south. The horizontal displacement of the western fault is 45 ft and that of the eastern fault is 85 ft. The two faults cut the stratigraphic section into three blocks of identical stratigraphy. Bedding in the central block is overturned and is perpendicular to the strike of bedding in the blocks on either side. These three blocks are cut by numerous minor faults.

The nodular limestone (Sm_6) of the "Red Band" (Fig. 23) is cut by several small thrust faults that are parallel to the strike of the bed and dip at a slightly steeper angle than the bed. The faults are 3 to 4 ft apart, and in one place there are as many as five of them. Since the bed is not competent, stress has been relieved along this layer by means of the thrusts. Sheets of nodular limestone have been thrust up and piled on one another, thus increasing the apparent thickness of the unit.

One outcrop of hornblende diorite 8 ft long by 6 ft wide is completely surrounded by rocks of the Moydart Formation. Contacts are well exposed and the relations are clearly visible. The block of hornblende diorite, bounded by five faults, appears to have been pushed forcefully up into the Moydart Formation. The rock is fractured and shattered, and some of the minerals have been altered to epidote and chlorite. The grains have been thoroughly shattered and the rock is now fine-grained and friable.

Summary of Structure at School Brook Cove

School Brook Cove is a large east-west fault zone. Most of the movement took place along two faults, one south of the beach and the other on the shore line. Movement along these two faults produced the mega-breccia between them. Before faulting, the general bedding attitude of these rocks (Ordovician quartzite, non-Arisaig Silurian, and Silurian Moydart Formation) was probably north-northeast with dips to the west.

Northern Boundary Fault of Cormorant Cliff Cove

Cormorant Cliff Cove is also an area where considerable faulting has taken place. The northern boundary of the zone is a fault that can be easily traced for over 600 ft. The fault is straight, and strikes due east with a vertical dip. The orientation of the shoreline at Cormorant Cliff Cove is controlled by the strike of the fault.

At the east end of the cove, extending eastward from an outcrop of Horton conglomerate that juts into the sea, there is a line of outcrops 50 to 75 ft from shore which rise above mean sea level. This line of partially submerged Horton conglomerate outcrops indicates the position of the fault offshore to the east. The Horton conglomerate is on the north side of the fault and Silurian and Devonian age rocks are on the south side. A 2-ft-wide interval of stretched conglomerate pebbles whose long axes are parallel to the strike of the fault lies within the fault zone.

Southern Boundary of Cormorant Cliff Cove

Outcrops of Horton Formation are found south of the Silurian rocks along the beach. They are presumed to be separated by an east-west striking fault.

Mega-Breccia of Cormorant Cliff Cove

Outcrops along the shore between the two faults are predominantly Silurian or Devonian (or both) Stonehouse Formation, and Devonian Knoydart Formation. Variously striking faults have cut the rock into large blocks and have produced a mega-breccia. Blocks of one type of Stonehouse Formation lithology are faulted against blocks of another Stonehouse Formation lithology. The breccia contains several blocks of Horton conglomerate. At the eastern end of the fault zone, there is one fault block of Ordovician quartzite, a block of hornblende diorite, and several blocks of Moydart Formation.

A large part of this disturbed zone is composed of red shale and sandstone of the Knoydart Formation that has been greatly sheared along cleavage planes. At one point, these red shales and sandstones comprise 125 ft of the width of the disturbed zone. Bedding features have been destroyed by the shearing and layers of various grain sizes have been converted into irregular patches. Streaked and lensoid areas of differing grain size give the appearance that the rock flowed plastically. The red shale and sandstone contains lenticular blocks, up to a foot long and 6 in. wide, of well-cemented white sandstone.

Foliation Cut by the Fault

The Knoydart red shale is cut by a cleavage which strikes N. 68° E. This cleavage is present in the shale adjacent to the fault but not in the sheared red shale caught in the fault plane. The angle between the strikes of the fault and the cleavage is 21°. Pebbles in the Horton conglomerate north of and to within 5 ft of the fault plane are partially stretched. The long axes of these pebbles are parallel to the strike of cleavage in the red shale. The cleavage or foliation is cut by the fault and is therefore older.

West End of Cormorant Cliff Cove Mega-breccia Zone

The western end of Cormorant Cliff Cove is bordered by the Horton Formation. The conglomerate is cut by many minor joints and faults, but the faulting has not produced any mega-breccia, or changed the attitude of bedding. The rock appears to be relatively undisturbed. The dip of the beds is nearly vertical and the strike is parallel to the east-west fault. Several bedding planes within the conglomerate show considerable shearing and movement. The fault movement that produced the displacement to the east could also have produced displacement along these bedding planes. Since the attitude of beds is the same as that of the fault plane, and since the beds are zones of weakness, east-west faulting could have occurred in this direction without disturbing the attitude of the conglomerate beds.

DISCUSSION

Cape George is an area of complexly folded and faulted Carboniferous and Lower Paleozoic rocks. The major faults in the area are: the fault zones of School Brook Cove and Cormorant Cliff Cove, the fault that strikes east-west from Ballantyne Cove to Livingstone Cove, and the northeasterly striking fault found near the source of Wilkie Brook and at Marsh Cove. Movement on the Ballantyne-Livingstone Cove Fault has thrust the Horton conglomerate, found to the north, over the McAras Brook flows and conglomerates found to the south. The movement along most of the faults of Cape George is post-Windsor. The intensive faulting throughout the northern end of the Cape is probably associated with movement or stress that accompanied the Hollow Fault, which parallels the northwestern coast of the Cape from Malignant Cove to Livingstone Cove.

The Carboniferous rocks consist of conglomerate, arkose, basalt flows, and sandstone assigned to the Mississippian Windsor, McAras Brook, and Horton Formations. The Windsor Formation rests with pronounced angular unconformity upon the Horton Formation, the latter being dated by the presence of *Lepidodendropsis*.

The Carboniferous sediments were deposited in a small basin area. During lower Horton time, there was considerable relief surrounding the basin, particularly to the north. Conglomerates were rapidly deposited in great thickness, not far from their source. Units of the Horton Formation have small areal extent, and the thicknesses of the units are large when compared to their extent.

The Horton Formation undergoes a facies change from boulder conglomerate to pebble conglomerate within a distance of 400 ft and to sandstone within several hundred feet. Considerable interfingering of sediments occurred during Horton time. The McAras Brook conglomerate and flows were deposited sometime during the Carboniferous, probably between the time of deposition of the Horton conglomerate and Horton sandstone units. Limestones of the Windsor Formation were deposited in a shallow marine basin. The process of the filling of the basin is exemplified by the onlap of the Carboniferous sediments upon the rhyolite island located at the present headwaters of Wilkie Brook.

The Horton Formation has been observed to rest with angular unconformity upon the Lower Devonian Knoydart Formation at the west end of School Brook Cove. The relation of the Knoydart Formation to the underlying strata of the Stonehouse Formation and Silurian Moydart Formation is structurally complicated by faulting; presumably, however, these strata are conformable as is the case elsewhere in Antigonish County and in adjoining Pictou County. The stratigraphic relation of the Silurian *Scolithus* quartzite to the Moydart Formation is also unknown due to structural complications. The Middle or Upper Ordovician rocks have been extensively intruded and baked by medium-grained granitic rocks (hornblende diorite), presumably during the Taconic orogeny, and are inferred on this basis to have undergone deformation during the Taconic. The pre-Silurian Browns Mountain Group, in the southern part of the Cape George region, is overlain unconformably by the Horton Formation.

The structure of the lower Paleozoic rocks at the tip of Cape George in School Brook and Cormorant Cliff Coves can best be summarized as blocks in what might be termed a mega-fault breccia. The detailed maps of the two coves indicate the position and relation of the blocks in this breccia. It should be noted that the blocks in the breccia have not been thoroughly intermixed, and some stratigraphic continuity can be discerned. The many units that are in the faulted zone, and their distinctive lithology, have aided in determining the structure.

At the eastern end of School Brook Cove, the Horton Formation is in contact with a fault complex of Middle or Upper Ordovician strata and associated hornblende diorite intrusives. Adjacent to the fault, in a zone about 1 ft wide, the pebbles in the Horton conglomerate have been severely stretched.

Farther to the west in School Brook Cove, the various units of the *Scolithus* quartzite are present, but their contacts with the Ordovician rocks, hornblende diorite, and the other Silurian and Lower Devonian rocks are obliterated everywhere by a maze of faults. Each of these minor faults within the fault zone is characterized by a thin layer of fine-grained, grayish-white fault gouge which seldom exceeds a few inches in thickness.

At the western end of School Brook Cove, the rocks of the Moydart, Stonehouse, and Knoydart Formations are present. These units are present as portions of three large blocks. The eastern and western blocks of the three are still right-side up, whereas some of the strata in the middle block are overturned. The contents of all three blocks have been badly faulted in detail, but the original stratigraphic relations can still be recognized, including the presence of individual beds. At the west end of the Cove, the Horton Formation rests with angular unconformity upon the Knoydart Formation. An illustration of the complexity of faulting here can be seen in the small block of hornblende diorite near the west end of the Cove, which is in fault contact with the Silurian on all four sides.

The lower Paleozoic rocks of Cormorant Cliff Cove are bounded to the north, west, and south by conglomerate of the Horton Formation. Where the Horton Formation is adjacent to the lower Paleozoic rocks, it is invariably greatly deformed and its pebbles are stretched, as was the case at School Brook Cove.

It is of interest to note that none of the Ordovician or Silurian fossils in the lower Paleozoic rocks have been badly deformed, indicating that the rocks have not been pervasively sheared, but have been displaced along distinct fault planes.

Middle or Upper Ordovician fossils have been identified from the quartzite of School Brook Cove. This is the first Middle or Upper Ordovician fossiliferous locality reported from Nova Scotia. The only fossiliferous outcrops of the quartzite are in School Brook Cove. One outcrop from which no fossils have been found is in Cormorant Cliff Cove. No similar rock types have been found in the pre-Silurian Browns Mountain Group.

The "Red Band" (Fig. 23), that marks the top of the Moydart Formation, is present at School Brook Cove and indicates the contact between the Stonehouse and the Moydart Formations.

Within the Moydart Formation, at School Brook and Cormorant Cliff Coves, there are sandstones that contain fish fragments. The fish are dated as late Silurian (Ludlovian) in age and are important because they date the upper part of the Moydart Formation and are rare in the Appalachians.

The Silurian tauna found in the *Scolithus* quartzite (Fig. 22) of School Brook Cove is unlike the fauna of the Arisaig Group or the Silurian fauna of any other locality in Nova Scotia.

Regional Relations

The relations of the Silurian and Lower Devonian rocks of Nova Scotia to those of adjacent portions of the northern Appalachians have, until recently, been uncertain. The pre-Carboniferous grain in coastal Maine and New Brunswick is northeasterly, and in southern Newfoundland it is again roughly northeast, but in the intervening area of northern Nova Scotia (excluding Cape Breton Island) it is almost east-west. Recent work in the northern Appalachians, both published (Berry and Boucot, 1970) and unpublished, helps to resolve this situation. It is also possible to interpret the Paleozoic of the Maritimes in the light of plate tectonics (McKerrow and Ziegler, 1971). The existing relations fall easily into place if it is assumed that the Silurian and Lower Devonian lithofacies units follow the regional grain; and that the northern Appalachians have a salient (in Nova Scotia) and a recess (in New Brunswick), like many other mountain belts. Once this reconstruction is allowed, the problem of setting up reasonable paleogeographic relations (taking into account both physical and biological environmental factors) becomes relatively simple for Maine, New Brunswick, and Nova Scotia, but the detailed relations of Southern Newfoundland to Nova Scotia are not clear. Figure 24 (Eastport Belt of Figure 24 equals the Coastal Volcanic Belt of Berry and Boucot, 1970) indicates three belts which have consistently distinct facies throughout the Silurian and Lower Devonian. These form the basis for the following discussion.

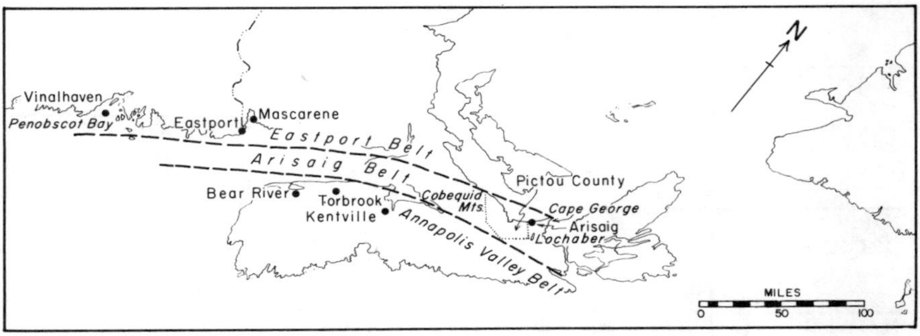

Figure 24. Northern Appalachian paleogeographic belts.

None of the pre-Silurian rocks of Antigonish and Pictou Counties can be dated well enough to make it worthwhile trying to relate them to the known Cambrian-Ordovician rock units of the Annapolis Valley and coastal New Brunswick regions. Beginning with the lower Llandovery, a reasonable picture can be established (Fig. 25, Table 13). The Beechhill Cove Formation of Arisaig contains a marine fauna of the same age as that in the Glencoe Brook Formation to the south in Pictou County. The Glencoe Brook is made up largely of quartzite lithologically similar to that found in the unfossiliferous White Rock Formation to the south in the Annapolis Valley belt. The White Rock is sandwiched between the Hali-

TABLE 13. Regional Correlations for

	ANNAPOLIS VALLEY, NOVA SCOTIA	SOUTHEAST PICTOU CO., NOVA SCOTIA
SIEGEN to LOWER GEDINNE	TORBROOK FORMATION (marine quartzite and shale)	KNOYDART FORMATION (nonmarine siltstone and shale)
PRIDOLI	NEW CANAAN FM. (marine volcanic breccia and slate) (near Wolfville only)	STONEHOUSE FORMATION (marine siltstone and shale)
LUDLOW	KENTVILLE FORMATION (marine shales)	KERROWGARE FORMATION (marine shales and some sandstone)
WENLOCK	WHITE ROCK FORMATION (quartzite and slate)	
LLANDOVERY		GLENCOE BROOK FM. (marine quartzite and mudstone)
?UPPER ORDOVICIAN	——	SUNNYBRAE FM. (shale and tuff) CHARCOAL FM. (tuff & conglomerate)
?LOWER ORDOVICIAN and TREMADOC	HALIFAX FORMATION (slates)	BROWNS MOUNTAIN GROUP (graywackes & slates)

fax and Kentville Formations, which contain graptolites respectively assigned to the Tremadoc (Smitheringale, 1960) and the Ludlow (Berry, 1958; Smitheringale, 1960; Crosby, 1962); its age is thus difficult to establish, but a correlation with the Glencoe Brook, at least in part, seems the most reasonable conclusion. In addition, the fauna of the Beechhill Cove Formation is much richer than the Glencoe Brook fauna; this, together with the greater abundance of the very shallow water brachiopod *Cryptothyrella* in the Glencoe Brook, suggests that the Glencoe Brook Formation was deposited closer to a (southern) shoreline. With such an interpretation in mind, the White Rock can be considered as a series of largely

Nova Scotia, New Brunswick, and Maine

<table>
<tr><th colspan="3">NORTHEAST PICTOU CO. AND ARISAIG, NOVA SCOTIA</th><th>COASTAL NEW BRUNSWICK AND MAINE</th></tr>
<tr><td colspan="3">KNOYDART FORMATION
(nonmarine siltstone and shale)</td><td>EASTPORT FORMATION
(volcanics and nonmarine sediments; basal marine)</td></tr>
<tr><td colspan="2">STONEHOUSE FORMATION
(marine siltstone and shale)</td><td rowspan="8">ARISAIG GROUP</td><td>PEMBROKE FORMATION
(volcanics and marine sediments)</td></tr>
<tr><td>Red member
(nonmarine marls and nodular 1st.)</td><td rowspan="2">MOY-DART FM.</td><td rowspan="3">EDMUNDS FORMATION
(volcanics, marine mudstones and siltstones)</td></tr>
<tr><td>Green member
(marine siltstone)</td></tr>
<tr><td colspan="2">McADAM BROOK FM.
(marine shale and siltstone)</td></tr>
<tr><td colspan="2">DOCTORS BROOK FM.
(marine shale and siltstone)</td><td rowspan="2">DENNYS FORMATION
(volcanics, marine clastics)</td></tr>
<tr><td colspan="2">FRENCH RIVER FM.
(marine siltstones and iron ore)</td></tr>
<tr><td colspan="2">ROSS BROOK FM.
(marine shales and siltstones)</td><td>QUODDY FORMATION
(volcanics, marine clastics)</td></tr>
<tr><td colspan="2">BEECHHILL COVE FM.
(marine siltstone)</td><td>——</td></tr>
<tr><td colspan="3">BEARS BROOK VOLCANIC GROUP
(volcanics and breccias)</td><td>——</td></tr>
<tr><td colspan="3">BROWNS MOUNTAIN GROUP
(graywackes & slates)</td><td>CHARLOTTE GROUP
(slates)</td></tr>
</table>

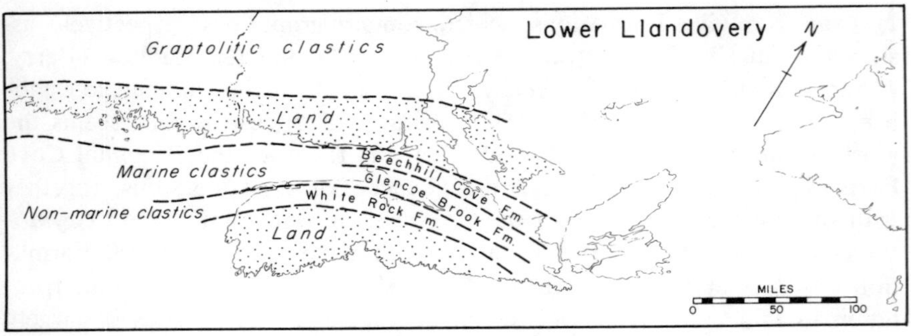

Figure 25. Lower Llandovery paleogeography.

nonmarine (to account for the lack of fossils) quartzites deposited adjacent to a land area which lay to the southeast of the Annapolis Valley. This area may have provided much of the quartz-rich debris in the Lower Silurian rocks to the north and northwest. Lower Silurian rocks earlier than C_4 (late upper Llandovery) are unknown in coastal Maine and New Brunswick (Berry and Boucot, 1970), and it may be that a second land area lay to the north and northwest of the Arisaig belt. The relative contributions and importance of these two potential source areas for the lower Llandovery sediments of Nova Scotia cannot be estimated at the present time. Studies of current indicators (Table 6) in the Beechhill Cove Formation suggest that the dominant local direction of transport was from the northeast. This direction can best be interpreted as parallel to both shorelines.

The middle and early upper Llandovery (pre-C_6) portion of the Arisaig Group (that is, the lower and middle members of the Ross Brook Formation and a lower portion of the Kerrowgare Formation) can best be interpreted as having been deposited in a similar linear basin to that prevailing in lower Llandovery times; although around C_{4-5} times, sedimentation and volcanic activity started in the Eastport belt to the northwest of the Arisaig belt. From C_6 through Wenlock time, the relation of the Arisaig Group (including the upper member of the Ross Brook Formation, the French River and Doctors Brook Formations, and the major portion of the Kerrowgare Formation) to the White Rock Formation is similar to that postulated for earlier Silurian times (Fig. 26).

The idea that the Arisaig Group is close to an emergent shoreline is reinforced by the occurrence in both middle and upper members of the Ross Brook Formation of an abundant representation of the *Eocoelia* Community, which indicates proximity to a shoreline (Ziegler, 1965). To the west and northwest, in coastal Maine and New Brunswick, the brachio-

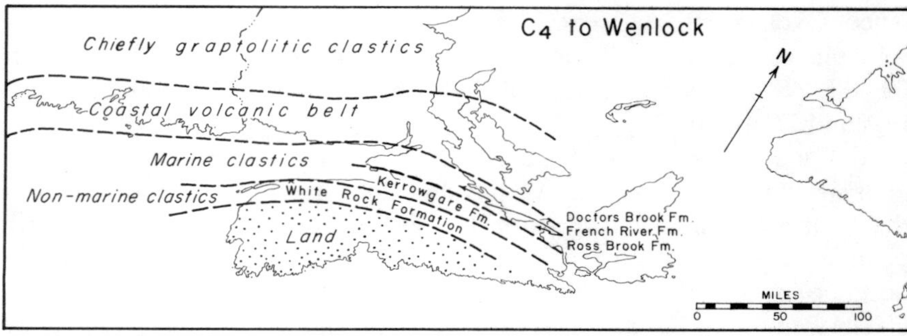

Figure 26. Upper Llandovery and Wenlock paleogeography.

pod communities are more variable; *Eocoelia, Pentamerus, Stricklandia,* and *Clorinda* Communities are all present in places (Berry and Boucot, 1970); Ziegler has found that the latter three invariably occur seaward of the *Eocoelia* Community. The sediments containing these latter communities are interbedded with volcanic rocks and the sea floor may have had a variable topography. At times the volcanics may have provided clastic contributions, and their volume relative to that gained from the White Rock Formation cannot be determined from the information at present available.

A major change in the paleogeography of the area occurred about the start of Ludlow times, when the nonmarine conditions to the south of the Arisaig belt gave way to graptolitic shales (Fig. 27). The McAdam Brook and Moydart Formations, and also the upper part of the Kerrowgare Formation, can be correlated with the graptolitic beds of the Kentville Formation (Berry, 1958; Smitheringale, 1960; Crosby, 1962) in the Annapolis Valley belt. At this time, coastal New Brunswick and Maine, in the Eastport belt, were still sites of marine deposition and contempo-

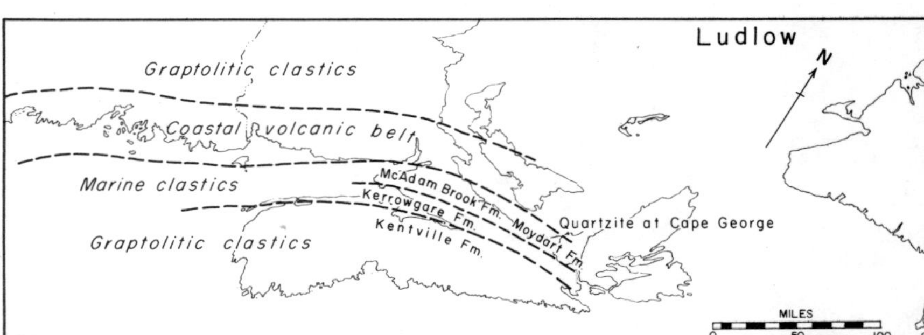

Figure 27. Ludlow paleogeography.

raneous volcanism. As there is now no land to the south of the Arisaig belt, the volcanics of the Eastport belt are probable sources of sediments for the Arisaig deposits; they include the igneous rocks present in the Edmunds Formation.

At Cape George, Nova Scotia, beds south of a major fault are similar to their time-equivalents at Arisaig, but contemporaneous beds north of this fault are quartzites with worm tubes. A region with coarser sediments, possibly a northeastward continuation of the Eastport belt, lay to the north of Arisaig during Ludlow times.

The Pembroke Formation is of Pridoli age and is equivalent to much of the Stonehouse Formation. The bulk of the overlying Eastport Formation is probably of nonmarine origin, as judged from its fauna which consists chiefly of pelecypods, gastropods, and ostracodes with almost no representation of wholly marine groups. Some of the lower Eastport is a lateral facies of the lower Gedinnian age part of the Stonehouse Formation (Fig. 28) and the overlying Knoydart, and could have been a potential source of sediment for them. To the south and southeast, in the Annapolis Valley belt, Stonehouse-Pembroke-Eastport time is represented by marine beds.

The nonmarine Knoydart Formation of Antigonish and Pictou Counties is best interpreted as a facies of the coeval marine beds occurring in the Annapolis Valley (Fig. 29). It is probable that a large portion of the Eastport Formation may also be coeval and of nonmarine origin, but the lack of fossils prevents any certainty upon this point. The nearest marine beds of Knoydart age to the north and northwest occur in central Maine and New Brunswick. These marine beds are too distant for their potential relations to the Knoydart to be inferred.

It is of importance to note that the rock units of the Arisaig Group, with the exception of the basal Beechhill Cove, have been found by Kelley (1966) and Harper (1964) in the Cobequid Mountains well to

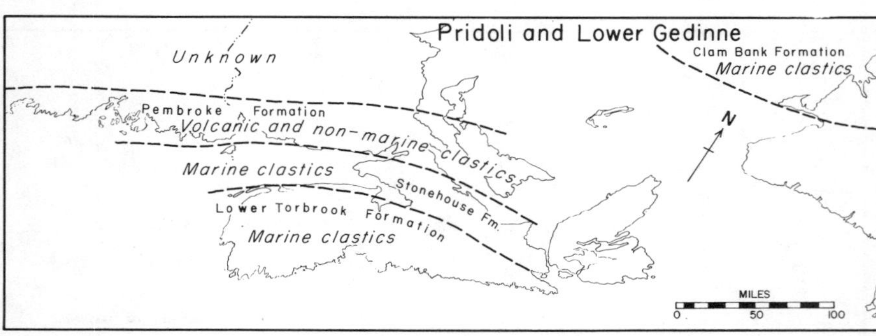

Figure 28. Pridoli and lower Gedinne paleogeography.

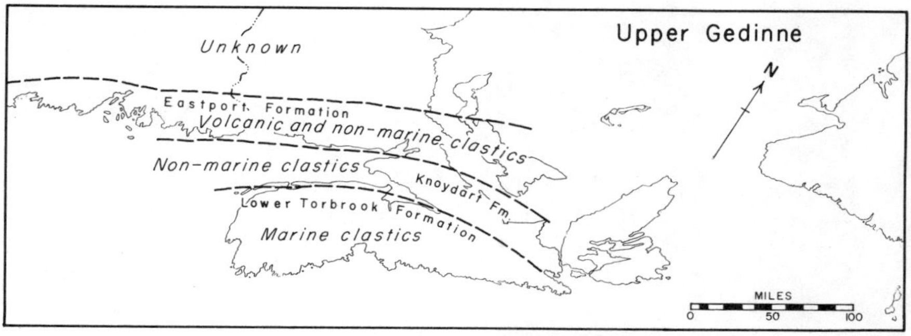

Figure 29. Upper Gedinne paleogeography.

the west of both Pictou and Antigonish Counties. Rocks of Arisaig Group age may occur, associated with volcanic rocks in the Mabou region of southwestern Cape Breton Island (Wait, 1959), but the data are too inconclusive to be used. The reputed Silurian age of the post-Early Ordovician, pre-Carboniferous Middle River Group (Weeks, 1954) of southeastern Cape Breton Island is too poorly established in the complete absence of paleontological information to be worthy of much consideration at this time. The northeast current indicators at Arisaig (Table 6) suggest, however, that this region to the northeast of Arisaig may also have made some contributions to the Arisaig sediments from time to time.

Acadian Orogeny

The evidence of major changes associated with the Devonian earth movements in the northern Appalachians may be examined under the following headings: (1) change from marine to nonmarine facies; (2) unconformities; (3) granites; (4) pebbles in younger conglomerates; (5) structural differences in rocks above and below the unconformities.

1. The changes from marine to nonmarine facies in Nova Scotia and the adjacent areas of the northern Appalachians are summarized in Figure 30. These occurred in the Eastport and in the Arisaig belts well before the onset of any widespread folding. The first change to nonmarine conditions occurred toward the end of Ludlow times in the Eastport belt and in the Arisaig belt. But this was temporary; marine conditions were again widespread during the Pridoli and the early part of the lower Gedinnian. Persistent nonmarine environments only appeared in the Eastport and Arisaig belts after the early Gedinnian. The sea, however, persisted in the southeasterly Annapolis Valley belt from the Ludlow through the lower Emsian. There is thus no direct time relation between the marine to nonmarine facies changes and the onset of folding in Devonian times.

2. Unconformities are present above Early Devonian (Siegenian or later) beds in the two northwesterly belts (where nonmarine conditions had prevailed for some time prior to the folding); the Annapolis Valley belt had marine sedimentation continuing until lower Emsian time.

After widespread folding throughout the northern Appalachians, the first postorogenic deposit in the region was the Upper Devonian Perry Sandstone, although still earlier (Middle Devonian) beds are present in a similar tectonic situation in central and northern Maine (Boucot and others, 1964, p. 93). Sedimentation was not resumed in the Nova Scotian Arisaig and Annapolis Valley belts until the Mississippian. The basal Horton conglomerates are generally unfossiliferous. They are normally overlain conformably by the Windsor Group of Viséan age, as, for example, at Arisaig where Windsor fossils are present on the shore 3,700 ft southwest of McAras Brook.

3. Many Devonian granites are present in the Eastport belt and in the Annapolis Valley belt, but Fairbairn and others (1960) report a Devonian

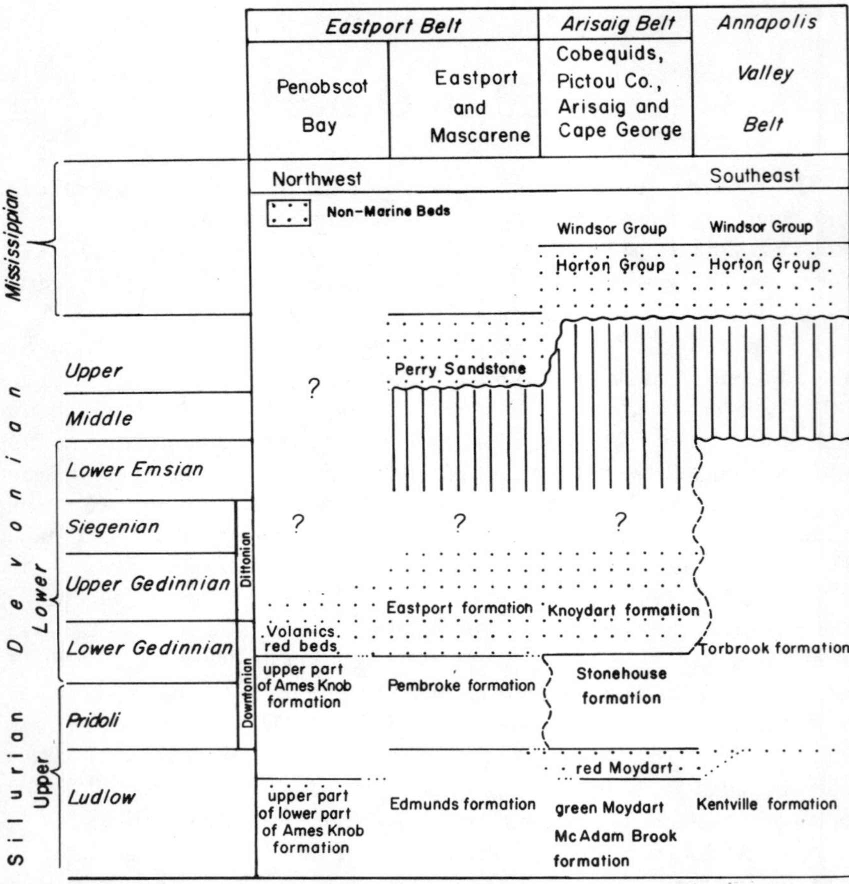

Figure 30. Stratigraphic relations of the Devonian beds of the northern Appalachians. Formations represented by dots are nonmarine.

age for a granite at James River, 7 mi west of Antigonish; this intrudes rocks of the Browns Mountain Group and may be an older granite which has undergone some changes during Devonian time. We conclude that there may be no Devonian granites in the Arisaig belt.

The granites in the Annapolis Valley belt have K/Ar and Rb/Sr ages of between 300 and 400 m.y.; two of these, which are known to cut Siegenian and lower Emsian beds of the Torbrook Formation at Bear River and Nictaux Falls, have ages of 363 to 370 m.y. (Fairbairn and others, 1960; Harland and others, 1964, p. 336-337).

In the Eastport belt, there is good stratigraphical control for the age of the St. George granite. It intrudes Silurian and Lower Devonian rocks;

boulders of this granite are common in the Upper Devonian Perry Formation (Bastin and Williams, 1914, p. 13; Boucot and others, 1964).

4. Pebbles in the Mississippian conglomerates are frequently polymict. This is true, for example, of the Horton conglomerate at School Brook Cove, Cape George, where the wide variety of Silurian and pre-Silurian sedimentary and igneous pebbles indicates major folding, uplift, and erosion in Devonian times. Other exposures of conglomerate on Cape George occur close to major faults, and their polymict nature could be entirely due to local uplift and erosion adjacent to the faults. In contrast, the Mississippian conglomerates of the McAras Brook Formation at Arisaig are monomict.

Along the shore near the mouth of McAras Brook (Pl. 1), for example, only pebbles of the Knoydart Formation are present in these beds. Many of the pebbles are argillaceous and show cleavage inherited from the parent rock from which they were derived. The major folding of the Knoydart thus took place before the Carboniferous.

On the shore southwest of McAras Brook, the current directions consistently have a component from the east; so it can be concluded that in early Mississippian times only Knoydart beds were exposed to the east of McAras Brook on both sides of the Hollow Fault (the Hollow Fault passes about 2 mi southeast of the mouth of McAras Brook). It is thus concluded that the pre-Horton folding and uplift was less than 1,000 ft (the thickness of the Knoydart Formation) at this point, otherwise Silurian pebbles would be included in the Mississippian conglomerates. It is also highly improbable that any vertical movements of more than 1,000 ft occurred on the Hollow Fault south of Arisaig during the Devonian. It should be noted, however, that the Horton Group (McAras Brook Formation) rests directly on the Stonehouse Formation at shore station 0 (Pl. 4); the eroded Stonehouse fragments were presumably transported to the west from this point, as they do not occur in the overlying conglomerates on the shore. In this area (the present coastline), the Knoydart was completely removed, indicating greater uplift and erosion in Devonian times to the west of the Arisaig area than in the region to the southeast of Arisaig.

5. The Mississippian rocks, which lie unconformably on the Knoydart and Stonehouse Formations in the McAras Brook region, have gently westerly dips; cleavage is absent. The strike of these beds is not very different from that of the underlying Lower Devonian beds. As elsewhere in Nova Scotia (Fyson, 1964, 1967), post-Carboniferous structures appear to follow pre-Carboniferous fold trends.

The other outcrop of Mississippian rocks in the Arisaig area is along the east branch of Doctors Brook (stations U89 to U103, Pl. 5) where sandstones and limestones of the McAras Brook and Ardness Formations occur in the Hollow Fault zone. These beds are just as deformed as the adjacent lower Paleozoic rocks, but they do not contain chlorite. It can be concluded that the deformation of the beds adjacent to the Hollow Fault at Arisaig occurred to a considerable degree in post-Carboniferous times, but that the chloritization of the lower Paleozoic beds took place before Mississippian times.

Appendix. Fossil Localities

All five-digit numbers refer to U.S. National Museum Locality Register.

Beechhill Cove Formation

Localities

10114 Shore, 2,050 ft northeast of McGillivray Brook, 150 ft north-northwest of overburden line. Plate 3.

10115 McGillivray Brook, 50 ft upstream from contact with underlying Bears Brook Volcanic Group. Plate 3.

10116 Shore, 2,175 ft northeast of McGillivray Brook, 60 ft north-northwest of overburden line. Plate 3.

10117 40 ft east of wagon road east of Doctors Brook, 275 ft north of old shore road. Plate 3.

10929 Arisaig Shore section. First point west of Dunn Point.

10930 North of Nova Scotia Route 45, 1,500 ft northeast of intersection of main road with Doctors Brook. Plate 2.

10952 West of Beechhill Cove.

	10114	10115	10116	10117	10929	10930	10952
Lingulid							X
Ptychopleurella? sp.				X			
Dolerorthis sp.				X			
Linoporella sp.		X		X			
Dalmanella primitiva			X				
D. (?) *primitiva*					X		
D. sp. A.		X					
D. sp. ?		X		X			
Isorthis sp. A.		X					
Mendacella arisaigensis	X	X		X		X	
Leptaena sp.			X	X			
Leptaenid				X			
Leptostrophia beechhillensis	X	X	X				
Fardenia? sp.		X	X				
"*Camarotoechia*" *bimesiornata*		X	X				
C. rossonia			X				
Cryptothyrella beechhillensis		X					
Gastropods			X				
Bivalves	X						
Trilobites			X	X			
Crinoids			X				

Ross Brook Formation, Upper Member

Localities

10120 Mill Brook, 90 ft south of Route 45. Plate 2.

10126 Shore 475 ft southwest of Arisaig Brook and 15 ft southeast of station 107. Plate 4.

10127 Shore 1,275 ft southwest of Arisaig Brook and 10 ft southwest of station 99. Plate 4.

10128 Shore 150 ft southwest of Smith Brook and 15 ft southwest of station 89. Plate 4.

10129 Shore 10 ft southwest of French Brook and 65 ft east of station 85. Plate 5.

10130 Doctors Brook, 240 ft northeast of station L12. Plate 5.

10131 South of Arisaig cemetery on east side of private road, and 400 ft southeast of Zeno MacDonald's garage. Plate 2.

10132 Doctors Brook West Branch, cliff exposure 210 ft west of station U51. Plate 5.

10923 Arisaig shore.

10932 West of Arisaig Brook.

10941 Doctors Brook, station M8½. Plate 5.

	10120	10126	10127	10128	10129	10130, 10131	10132	10923	10932	10941
Salopina conservatrix			X		X		X			
Visbyella nana	X		X							
Leptaena cf. *depressa*				X						
Pholidostrophia (Mesopholidostrophia) cf. *nitens*			X							
Protochonetes tenuistriatus		X	X	X	X			X		
"Camarotoechia" rossonia	X	X						X	X	
Plagiorhynchia aff. *glassi*	X	X		X				X	X	
Rhynchonelloids							X			
Atrypa cf. *reticularis*			X							
Eocoelia sulcata			X	X	X	X				
Hyattidina sp.			X	X						
Bivalves				X	X					
Monograptus clintonensis				X	X					
Unidentified fossils										X

Ross Brook Formation, Middle Member

Localities

10121 Shore, 950 ft northeast of Arisaig Brook, and 10 ft southeast of station 122. Plate 4.

10122 Shore, 475 ft northeast of Arisaig Brook, and 10 ft southwest of station 117. Plate 4.

10123 Shore, 50 ft southeast of station 112, and 20 ft northeast of Arisaig Brook. Plate 4.

10124 East of Arisaig Brook, 140 ft southeast of shore station 112, and 70 ft northeast of Brook station 3. Plate 7.

10125 Southeast of Nova Scotia Route 45, and 100 ft southwest of Mill Brook. Plate 2.

10857 McGillivray Brook, 400 ft north of Route 45. Plate 2.

10858 McGillivray Brook, 50 ft north of Route 45. Plate 2.

10859 McGillivray Brook, 650 ft south of Route 45. Plate 2.

	10121	10122	10123	10124	10125	10857	10858	10859
Lingulid						X		
Visbyella nana				X				
V. nana?	X							
"*Camarotoechia*" *rossonia*				X				
Plagiorhynchia aff. *glassi*	X			X				
Rhynchonellid	X						X	
Eocoelia hemisphaerica	X							
E. intermedia							X	
E. curtisi		X						X
Platyceratid gastropod							X	
Retiolites geinitzianus var. *augustidens*			X					
Monograptus cf. *marri*			X					
M. sp. (?*M. marri*)					X			

Ross Brook Formation, Lower Member

Localities

10118 Shore, 2,625 ft northeast of McGillivray Brook, 150 ft north-northwest of overburden line. Plate 3.

10119 Shore, 2,025 ft northeast of McGillivray Brook, 5 ft north-northwest of overburden line. Plate 3.

	10118	10119
Pentlandina? sp.	X	
Orthograptus bellulus		X
Climacograptus scalaris var. *normalis*		X
C. rectangulatus		X
Monograptus acinaces		X
M. lobiferus		X

French River Formation

Localities

10860 Doctors Brook, 175 ft northwest of junction of east and west branches. Plate 2.
10861 McGillivray Brook, 1,300 ft south of Route 45. Plate 2.
10862 Hillside, 2,600 ft southeast of mouth of McGillivray Brook. Plate 2.
10133 Doctors Brook, 25 ft southwest of station M25. Plate 5.
10134 Doctors Brook, 30 ft southwest of station L39. Plate 5.
10135 Doctors Brook, 15 ft south of station M28. Plate 5.
10136 Doctors Brook, 45 ft southwest of station M28. Plate 5.
10137 Doctors Brook, 60 ft southwest of station M28. Plate 5.
10138 Doctors Brook, 15 ft north of station M27. Plate 5.
10139 Doctors Brook, 120 ft south of station M27. Plate 5.
10140 Doctors Brook, 5 ft north of station M26. Plate 5.
10141 Doctors Brook, 40 ft northwest of station M14. Plate 5.
10142 Doctors Brook, 50 ft northwest of station M11. Plate 5.
10143 Doctors Brook, 50 ft southwest of station M9. Plate 5.
10144 Doctors Brook, West Branch 10 ft north of station U51. Plate 5.
10145 Doctors Brook, West Branch 30 ft east of station U49. Plate 5.
10146 Doctors Brook, West Branch 40 ft southwest of station U49. Plate 5.
10147 Doctors Brook, West Branch 25 ft northeast of station U48. Plate 5.
10148 Doctors Brook, West Branch 15 ft north of station U43. Plate 5.
10149 Beechhill Cove Brook, 950 ft southeast of Route 45. Plate 2.
10150 McGillivray Brook, 1,150 ft southeast of Route 45 and 600 ft south-southwest of barn. Plate 2.
10151 Mill Brook, 1,400 ft south of Route 45. Plate 2.
10152 Ross Brook, South Branch 30 ft northwest of station 65. Plate 6.
10153 Ross Brook, South Branch 25 ft south of station 68. Plate 6.
10154 Ross Brook, South Branch 10 ft east of station 69. Plate 6.
10155 Ross Brook, South Branch 5 ft south of station 70. Plate 6.
10156 Ross Brook, North Branch 25 ft northeast of station 113. Plate 6.
10157 Slope southwest of Macdonald Road, Arisaig Brook, 10 ft east of station 66. 70 ft stratigraphically below iron ore. Plate 7.
10158 Iron ore mine northeast of Arisaig Brook, 70 ft north of station 69. Plate 7.
10159 Slope northeast of Smith Brook 70 ft northwest of bridge on old shore road. Plate 4.
10160 Slope northeast of French Brook 100 ft north-northwest of bridge on old shore road. Plate 4.
10911 Middle Doctors Brook, Arisaig.
10912 Lower part of Doctors Brook.
10924 Doctors Brook (lower part).
10927 Doctors Brook, station M27½. Plate 5.

French River Formation (continued)

Localities

10931 Doctors Brook, station M1. Plate 5.
10933 Doctors Brook.
10934 Beechhill Cove Brook, 650 ft south of Route 45. Plate 2.
10936 Doctors Brook, station M25. Plate 5.
10939 Middle Doctors Brook, 450 ft northeast of station M14. Plate 5.
10940 Doctors Brook, station L38½. Plate 5.
10947 Middle Doctors Brook, 275 ft north of junction of east and west branches. Plate 5.
10951 Doctors Brook, West Branch station U49. Plate 5.

? French River Formation

Localities

10917 Middle Doctors Brook, 75 ft downstream from station MO. Plate 5.
10921 Middle Doctors Brook.
10938 Doctors Brook.
10944 Doctors Brook.

N.B. 10921, 10938, and 10944 have no precise localities.

	10917	10921	10938	10944
Fardenia sp?	X	X		
Davidsoniid			X	
Unidentified fossils				X

French River Formation	10860	10861	10157	10154	10862	10151	10133	10134	10141	10143	10135	10127	10136	10137	10138	10139
Linguloids																X
Salopina submedia		X													X	X
S. submedia?	X															
Idiorthis matura																
Isorthis macadamensis											X		X		X	
I. sp.																
Resserella cf. *R. concavoconvexa*																
Dalejina sp.																
Heterorthella maehli																
Marklandella macadamica	X															
Leptaena cf. *depressa*																
Leptostrophia sp.																
Amphistrophia aff. *prolongata*																
Mclearnites sp.																
Fardenia sp.																
Protochonetes novascoticus																
Sphaerirhynchia sp.									X					X	X	
"Camarotoechia" aff. *planorugosa*																
Plagiorhynchia cf. *plastica*																X
Rhynchonelloids	X												X			
Atrypa cf. *reticularis*													X		X	
Rhynchospirina sp. A.																
Meristina billingsi		X		X							X		X	X	X	
Striispirifer stonehousensis																X
Monograptus distans							X									
Trilobites																
Unidentified fragments of brachiopods																
Unidentified fossils																

10860 10861 10157 10154 10862 10151 10133 10134 10141 10143 10135 10127 10136 10137 10138 10139

10140																					X								X	10140
10142							X												X						X			X		10142
10144 / 10931																												X		10144 / 10931
10145							X			X			X				X		X										X	10145
10146													X						X							X				10146
10147													X															X		10147
10148							X															X			X			X		10148
10149						X															X									10149
10150							X			X		X		X	X		X	X	X					X				X		10150
10152							X				X																			10152
10153							X					X																		10153
10155							X		X						X			X					X							10155
10156				X		X				X	X																	X		10156
10158							X	X			X	X		X		X		X	X			X						X		10158
10159											X	X																		10159
10160											X																	X		10160
10924 / 10933 / 10934											X																			10924 / 10933 / 10934
10911																			X											10911
10912							X		X	X					X				X						X				X	10912
10936										X																				10936
10939	X																													10939
10947	X																													10947
10951	X																													10951
10940		X																												10940

Doctors Brook Formation

Localities

10161 Doctors Brook, East Branch 15 ft southwest of station U72. Plate 5.
10162 Doctors Brook, West Branch 30 ft northeast of station U54. Plate 5.
10163 Doctors Brook, West Branch 30 ft north of station U53. Plate 5.
10164 Shore 440 ft southwest of French Brook and 15 ft east of station 31. Plate 4.
10165 Shore 1,450 ft northeast of McAdam Brook and 25 ft northeast of station 64. Plate 4.
10909 Shore, Arisaig.
10913 Doctors Brook, East Branch, station U79. Plate 5.
10914 Doctors Brook, East Branch, station U64. Plate 5.
10915 Doctors Brook, West Branch, station U50. Plate 5.
10916 Doctors Brook, East Branch, station U64. Plate 5.
10918 Doctors Brook, East Branch, 75 ft up tributary, southeast of station U71. Plate 5.
10919 Doctors Brook, station U64. Plate 5.
10942 Doctors Brook, West Branch, station U50. Plate 5.
10945 Doctors Brook, West Branch, station U53. Plate 5.
10950 Doctors Brook, West Branch, station U50. Plate 5.

	10161	10162	10163	10164	10165	10909	10913, 10914	10915, 10916	10918, 10919	10942, 10945, 10950
Linguloids					X					
Salopina submedia				X	X	X				
Idiorthis matura				X		X				
Idiorthis matura?			X							
Isorthis macadamensis						X				
Marklandella macadamica				X						
Leptaena cf. *depressa*				X		X				
Leptostrophia sp.				X		X				
Amphistrophia aff. *prolongata*				X						
A. cf. *striata*						X				
Fardenia sp.	X									
Fardenia sp?								X		
Protochonetes novascoticus				X		X				
Sphaerirhynchia sp.		X	X	X		X				
"*Camarotoechia*" aff. *planorugosa*				X	X	X				
Rhychonelloids						X				
Atrypa cf. *reticularis*				X						

Doctors Brook Formation (continued)

	10161	10162	10163	10164	10165	10909	10913, 10914	10915, 10916	10918, 10919	10942, 10945, 10950
Howellella moydartensis						X				
Bivalves				X	X	X				
Gastropoda					X					
Cephalopoda					X	X				
Unidentified fossils										X
Trilobites						X				

McAdam Brook Formation

Localities

Upper Member:

10170 Shore 990 ft southwest of McAdam Brook and 15 ft northeast of station 39. Plate 4.

Lower Member:

10166 Southeast side of old shore road, 90 ft southwest of French Brook. Plate 4.

10167 25 ft southwest of Smith Brook, 150 ft north of old shore road. Plate 4.

10168 Shore 35 ft east of McAdam Brook and 45 ft south of station 50. Plates 4 and 8.

10169 Shore 650 southwest of McAdam Brook and 50 ft south of station 43. Plate 4.

	Upper Member: 10170	Lower Member: 10166	10167	10168	10169
Salopina submedia	X		X	X	X
Fardenia sp.		X			
Protochonetes novascoticus	X			X	X
Sphaerirhynchia sp.			X	X	X
"*Camarotoechia*" aff. *planorugosa*	X			X	X
Plagiorhynchia sp. A.			X		X
Howellella moydartensis			X		
Monograptus varians				X	
Plectonotus		X			
Bivalves	X		X	X	X
Cephalopods	X				X
Bryozoa	X				

Moydart Formation, Green Member

Localities

10171 Slope northeast of Smith Brook, 320 ft southeast of old shore road bridge. Plate 4.

10172 Slope northeast of French Brook, 380 ft southeast of old shore road bridge. Plate 4.

10173 Smith Brook 380 ft north-northwest of Route 45 bridge. Plate 4.

10174 Smith Brook 280 ft north-northwest of Route 45 bridge. Plate 4.

Moydart Formation (continued)

Localities

10175 Smith Brook 190 ft north of Route 45 bridge. Plate 4.
10176 Smith Brook 110 ft north-northwest of Route 45 bridge. Plate 4.
10177 Smith Brook 100 ft north of Route 45 bridge. Plate 4.
10178 North of Red Moydart on east of Route 45, 2,200 ft southwest of bridge over Smith Brook. Plate 2.
10179 Upper Smith Brook, 2,800 ft southeast of Route 45 bridge. Plate 2.
10180 Small ridge in scrub, 1,000 ft east-southeast of mouth of McAdam Brook. Plate 2.
10181 Shore 1,380 ft southwest of McAdam Brook and 30 ft southeast of station 35. Plate 4.
10182 Shore at Moydart Point and at station 27. Plate 4.
10183 Shore 270 ft west of McDonald Brook and 5 ft east of station 21. Plate 4.
10908 Arisaig shore.
10920 McAdam Brook, station L20. Plate 8.
10935 French Brook between old and new roads, about 600 ft from the shore.

	10171	10172	10173	10174	10175	10176	10178	10179	10177	10180	10181	10182	10183	10908	10920	10935
Linguloids											X		X			
Salopina submedia		X	X	X					X		X	X	X	X		
Shaleria gilpini				X								X	X			
Lissostrophia sp.												X				
Orthotetacid				X								X				
Protochonetes novascoticus		X		X			X		X		X	X	X	X		
Sphaerirhynchia saffordi		X										X		X		X
Eatonioides lamellornatus													X	X		
"Camarotoechia" aff. *planorugosa*		X	X		X							X	X	X		
"C." squamifera				X	X					X	X	X		X		
"C." sp.	X														X	
Rhynchonelloids																X
Rhynchospirina acadiae			X						X				X			
Hyattidina northumberlandensis				X	X				X	X	X	X				
Delthyris (Quadrifarius) rugaecostatus?												X				
Howellella moydartensis		X		X							X			X		X
Bivalves			X						X			X	X			
Gastropods				X	X						X					
Trilobites				X							X	X	X			

Stonehouse Formation

Localities

10184 Shore 440 ft southwest of McDonald Brook and 30 ft northeast of station 19. Plate 4.
10185 Shore 600 ft southwest of McDonald Brook and 25 ft west-southwest of station 18. Plate 4.
10186 On slope 490 ft south-southwest of mouth of McDonald Brook and 260 ft southeast from shoreline at station 19. Plate 4.
10187 Shore 1,000 ft southwest of McDonald Brook and 50 ft south-southwest of station 14. Plate 4.
10188 Shore 40 ft northeast of station 11. Plate 4.
10189 Shore 55 ft southwest of station 8. Plate 4.
10190 Shore 45 ft northeast of station 4. Plate 4.
10191 Shore 20 ft south of station 4. Plate 4.
10192 Shore 60 ft east of station 0. Plate 4.
10193 In brook, 730 ft southeast (that is, upstream from mouth) of shore station 8. Plate 4.
10194 On slope 770 ft east-northeast of shore station 8, and 550 ft southeast of shoreline at station 15. Plate 4.
10195 McDonald Brook, 810 ft southeast of (that is, upstream from) mouth. Plate 4.
10196 On ridge 720 ft east of mouth of McDonald Brook. Plate 2.
10197 On hill 1,100 ft south-southwest of mouth of McAdam Brook. Plate 2.
10198 On hill 750 ft south of mouth of McAdam Brook. Plate 2.
10199 On slope east of McAdam Brook, 90 ft east of station L13.
10200 McAdam Brook, 10 ft north of station L12. Plate 8.
10201 McAdam Brook, 25 ft north of station L11. Plate 8.
10202 McAdam Brook, station L9. Plate 8.
10203 McAdam Brook, East tributary, station 10C. Plate 8.
10204 McAdam Brook, East tributary, station 10J. Plate 8.
10205 McAdam Brook, station U14. Plate 8.
10206 McAdam Brook, 50 ft east of station U15. Plate 8.
10207 McAdam Brook, 15 ft north of station U41. Plate 8.
10208 West of Route 45, 2,500 ft southwest of bridge over Smith Brook. Plate 2.
10209 Upper Smith Brook, 3,125 ft south-southwest of Route 45 bridge. Plate 2.
10210 Arisaig Brook, 65 ft north of station 100. Plate 7.
10211 Arisaig Brook, 10 ft northwest of station 107. Plate 7.
10212 Arisaig Brook, 30 ft southeast of station 109. Plate 7.
10213 On slope 810 ft northeast of 4th bridge in Arisaig Brook (that is, bridge 375 ft south of station 109). Plate 2.
10214 2,750 ft east-northeast of Arisaig Brook station 96 (Red Moydart). Plate 2.

Stonehouse Formation (continued)

Localities

10910 Arisaig shore section.
10925 Between Arisaig Brook and McAdam Brook approx. 1 mi south of Route 45. Plate 2.
10926 McAdam Brook station L11. Plate 8.
10928 2,400 ft east of Arisaig Brook and about 4,000 ft south of Route 45. Plate 2.
10937 McAdam Brook.

Stonehouse Formation	10184	10185	10186	10187	10188	10189	10190	10191	10192	10193	10194	10195	10196
Linguloid	X	X		X	X				X				
Salopina missendenensis				X				X	X				
Salopina submedia													
Schizophoria sp.				X	X								
Isorthis (Protocortezorthis) fornicatimcurvata	X	X	X	X	X	X	X			X	X	X	X
Leptaena aff. *rhomboidalis*	X	X											
Shaleria gilpini		X			X		X	X	X	X	X	X	
Protochonetes novascoticus						X	X	X	X	X	X	X	
Protochonetes stonehousensis	X	X		X	X				X				
Mclearnites mertoni	X	X		X									
"*Camarotoechia*" *planorugosa*	X	X	X	X	X	X	X	X	X				X
Atrypa cf. *gedinniana*				X									
Rhynchospirina sinuata	X	X	X	X			X	X	X	X			
R. sp. A.		X											
Protathyris? sp.				X									
Delthyris (Quadrifarius) rugaecosta	X	X	X	X	X		X	X	X	X		X	X
Howellella moydartensis?		X											
Bryozoa		X	X	X									
Bivalves		X		X		X		X	X				
Cephalopods		X				X			X				
Gastropods						X			X				
Crinoids													
Trilobites	X			X	X	X	X	X	X				
Ostracoda		X						X	X				
Tentaculites				X									
	10184	10185	10186	10187	10188	10189	10190	10191	10192	10193	10194	10195	10196

10197						X								X				10197
10198						X		X		X		X		X				10198
10199					X	X				X		X	X	X				10199
10200						X												10200
10201						X		X		X		X	X	X		X		10201
10202			X	X	X		X			X	X			X		X		10202
10203		X			X	X				X		X	X	X		X		10203
10204			X			X		X				X	X	X				10204
10205		X	X	X	X					X		X		X		X	X	10205
10206		X	X	X	X			X		X		X		X		X	X	10206
10208														X				10208
10209												X						10209
10210						X				X			X	X		X		10210
10211			X			X		X	X	X		X	X	X		X		10211
10212												X	X		X	X		10212
10213					X	X						X						10213
10214						X				X			X					10214
10910						X		X		X		X		X				10910
10925					X							X			X			10925
10928						X							X	X			X	10928
10937	X																	10937
10922					X	X				X		X		X		X		10922
10926												X			X			10926
11336										X								11336

General References

Aldermann, S. S. Jr., 1948, The Lower Horton Conglomerate of Cape George, Nova Scotia [B.S. thesis]: Cambridge, Massachusetts Inst. Technology.

Allen, J.R.L., 1960, Cornstone: Geol. Mag., v. 97, p. 43–48.

——1961, Sandstone-plugged pipes in the Lower Old Red Sandstone of Shropshire, England: Jour. Sed. Petrology, v. 31, p. 325–335.

——1963, Depositional features of Dittonian rocks: Pembrokeshire compared with the Welsh Borderland: Geol. Mag., v. 100, p. 385–400.

Allen, J.R.L., and Tarlo, L. B., 1963, The Downtonian and Dittonian facies of the Welsh Borderland: Geol. Mag., v. 100, p. 120–155.

Ami, H. M., 1895, Notes on a collection of Silurian fossils from Cape George, Antigonish County, Nova Scotia, with descriptions of four new species: Nova Scotia Inst. Nat. Sci., v. VIII, p. 411–415.

——1900, Synopsis of the geology of Canada: Royal Soc. Canada Trans., v. VI, 2d Ser., sec. 4, p. 187–225.

——1901, Knoydart Formation of Nova Scotia: Geol. Soc. America Bull., v. XII, p. 301–312.

——1902, Description of tracks from the fine-grained siliceous mudstones of the Knoydart Formation (Eo-Devonian) of Antigonish County, Nova Scotia: Nova Scotia Inst. Nat. Sci., v. X, pt. 3, p. 330–332.

Ball, H. W., and Dineley, D. L., 1961, The Old Red Sandstone of Brown Clee Hill and adjacent area. I. Stratigraphy: British Mus. (Nat. History) Bull., Geology 5, p. 178–242.

Bastin, E. S., and Williams, H. S., 1914, Description of the Eastport Quadrangle: U.S. Geol. Survey Geol. Atlas, folio 192.

Beavon, R. V., Fitch, F. J., and Rast, N., 1961, Nomenclature and diagnostic characters of ignimbrites with reference to Snowdonia: Geol. Jour., v. 2, p. 600–611.

Bell, W. A., 1929, Horton – Windsor District Nova Scotia: Geol. Survey Canada Mem. 155, 268 p.

——1960, Mississippian Horton Group of type Windsor – Horton District, Nova Scotia: Geol. Survey Canada Mem. 314, 58 p.

Benson, D. G., 1964, Lochaber map-area: Canada Geol. Survey, Map 58–1963.

——1970a, Notes to accompany geological maps of Antigonish and Cape George map-areas, Nova Scotia: Canada Geol. Survey Paper 70–8, 2 p.

——1970b, Notes to accompany maps of the geology of Merigomish and Malignant Cove map-areas, Nova Scotia: Canada Geol. Survey Paper 70–9, 4 p.

Berry, W.B.N., 1958, Late Silurian graptolites in North America [abs.]: Geol. Soc. America Bull., v. 69, p. 1534.

Berry, W.B.N., and Boucot, A. J., 1970, Correlation of the North American Silurian rocks: Geol. Soc. America Spec. Paper 102, 289 p.

Boucot, A. J., 1960, Lower Gedinnian brachiopods of Belgium: Memoires de l'Institut Geol., Univ. de Louvain, tome 21, p. 201–324.

——1961, Appalachian Siluro-Devonian, *in* Some aspects of the Variscan fold belt: Manchester, England, Manchester Univ. Press, p. 155–163.

Boucot, A. J., Fletcher, R. and Griffin, J. R., 1959, Middle or Upper Ordovician in Nova Scotia [abs.]: Geol. Soc. America Bull., v. 70, no. 12, pt. 2, p. 1572.

Boucot, A. J., Field, M. T., Fletcher, R., Forbes, W. H., Naylor, R. S., and Pavlides, L., 1964, Reconnaissance bedrock geology of the Presque Isle Quadrangle, Maine: Maine Geol. Survey Quad. Mapping Ser. 2, p. 1–123.

Boucot, A. J., Johnson, J. G., Harper, C. W., Jr., and Walmsley, V. G., 1966, Silurian brachiopods and gastropods of southern New Brunswick: Geol. Survey Canada Bull., v. 140, 45 p., 18 pls.

Copeland, M. J., 1964, Canadian fossils, Ostracoda: Some Silurian species: Canada Geol. Survey Bull. 117, p. 1–20.

Crosby, D. G., 1962, Wolfville map-area, Nova Scotia (21 H 1): Canada Geol. Survey Mem. 325, p. 1–67.

Daly, R. A., 1901, The physiography of Acadia: Mus. Comp. Zool. Bull. 38, p. 73–103.

Dawson, J. W., 1845, On the Lower Carboniferous rocks, or gypsiferous formation of Nova Scotia, with geological map of Nova Scotia by Abraham Gesner: Geol. Soc. London Quart. Jour., v. 1, p. 26–34.

——1850, On the metamorphic and metalliferous rocks of eastern Nova Scotia: Geol. Soc. London Quart. Jour., v. 6, p. 347–364.

——1855, Acadian geology (1st ed.): Edinburgh.

——1868, Acadian geology (2d ed.): London.

——1878, Acadian geology (3d ed.): London.

——1881, New facts respecting the geological relations and fossil remains of the Silurian iron ores of Pictou, Nova Scotia: Canada Nat., n. s., v. 9, p. 313–314, 338.

——1889, Hand book of Acadian geology: Montreal.

——1891, The geology of Nova Scotia, New Brunswick, and Prince Edward Island, or Acadian geology (4th ed.): London, 694 p. (102 p., 37 p. supps).

Decker, R. W., 1949, Study of the Upper Horton sediments in Northern Nova Scotia and on Cape Breton Island: Annual Report on Mines, Province of Nova Scotia, p. 128–167.

Denison, R. H., 1955, Early Devonian vertebrates from the Knoydart Formation of Nova Scotia: Fieldiana—Zoology, v. 37, p. 449–464.

——1956, A review of the habitat of the earliest vertebrates: Fieldiana—Geology, v. 11, p. 361–457.

Dewey, J. F., 1963, The lower Paleozoic stratigraphy of central Murrisk, County Mayo, Ireland, and the evolution of the South Mayo Trough: Geol. Soc. London Quart. Jour., v. 119, p. 313–344.

——1965, The nature and origin of kink-bands: Tectonophysics, v. 1 (6), p. 459–494.

Dewey, J. F., and McManus, J., 1964, Superposed folding in the Silurian rocks of County Mayo, Eire: Geol. Jour., v. 4, p. 61–76.

Dineley, D. L., 1962, A *Climatius*-like spine from the Stonehouse Formation, Upper Silurian, Nova Scotia: Canada Nat. Mus. Nat. History Paper, Nr. 16, 4 p.

——1963, The "Red Stratum" of the Silurian Arisaig Series, Nova Scotia: Canada Jour. Geology, v. 71, no. 3, p. 523–524.

——1964, New specimens of *Traquairaspis* from Canada: Paleontology, v. 7, pt. 2, p. 210–219.

——1967, The Lower Devonian Knoydart faunas: Linnean Soc. London Jour. Zoology, v. 47, 311, p. 15–29.

Donath, F. A., 1964, Folds and folding: Geol. Soc. America Bull., v. 75, p. 45–62.

Erhart, H., 1956, La vie vegetale continentale aux epoques pre-Devoniennes vue sous l'angle de la theoree bio-rhexistasique et des derrieres decouvertes palynologiques: Soc. Géol. France Bull., v. 6, p. 445–450.

Erlanger, George, and Howard, Viet, 1959, Report on geologic reconnaissance of Cape George area, Nova Scotia: Massachusetts Inst. Technology Geol. Summer Camp Rept. [unpub.].

Fairbairn, H. W., Hurley, P. M., Pinson, W. H., and Cormier, R. F., 1960, Age of the granitic rocks of Nova Scotia: Geol. Soc. America Bull., v. 71, p. 399–413.

Fletcher, Hugh, 1887, Report on geological surveys and explorations in the counties of Guysborough, Antigonish, and Pictou, Nova Scotia: Canada Geol. Survey Ann. Rept. for 1886, pt. P, v. II, p. 5–128P.

——1893, Report on geological surveys and explorations in the counties of Pictou and Colchester, Nova Scotia: Canada Geol. Survey Ann. Rept. for 1889–1891, v. V, p. 5–193.

Fleuty, M. J., 1964, The description of folds: Geol. Assoc. Canada Proc., v. 75, p. 461–492.

Fyson, W. K., 1964, Repeated trends of folds and cross-folds in Paleozoic rocks, Parrsboro, Nova Scotia: Canada Jour. Earth Sci., v. 1, p. 167-183.

——1966, Structures in the lower Paleozoic Meguma Group, Nova Scotia: Geol. Soc. America Bull., v. 77, p. 931–944.

——1967, Gravity sliding and cross-folding in Carboniferous rocks, Nova Scotia: Am. Jour. Sci., v. 205 (1), p. 1–11.

Gesner, A., 1836, Remarks on the geology and mineralogy of Nova Scotia: Halifax, 272 p.

Goldthwait, J. W., 1924, Physiography of Nova Scotia: Canada Geol. Survey Mem. 140, p. 1–179.

Hall, J., 1860, Silurian fossils of Nova Scotia: Canada Nat. and Geology, v. 5, p. 144–159.

Harland, W. B., Smith, A. G., Wilcock, B., 1964, The Phanerozoic time-scale: Geol. Soc. London Quart. Jour., v. 120S, p. 1–458.

Harper, C. W., 1964, The brachiopod fauna of the Arisaig Series (Silurian-Lower Devonian) of Nova Scotia [Ph.D. thesis]: Pasadena, California Inst. Technology, p. 1–454.

Harper, C. W., Boucot, A. J., and Walmsley, V. G., 1969, The rhipidomellid brachiopod subfamilies *Heterorthinae* and *Platyorthinae* (New): Jour. Paleontology, v. 43, no. 1, p. 74–92, Pls. 15-17, 2 figs.

Hartt, C. Fred, 1868, On the subdivision of the Acadian Carboniferous limestones with a description of a section across these rocks at Windsor, Nova Scotia: Canada Nat. Ser., v. III, p. 212–224.

Hills, E. S., 1963, Elements of structural geology: London, Methuen, 483 p.

Honeyman, Rev. D., 1859, On the fossiliferous rocks of Arisaig: Lit. and Sci. Soc., Halifax, Trans., p. 19–29.

——1864, On the geology of Arisaig, Nova Scotia: Geol. Soc. London Quart. Jour., v. 20, p. 333–345.

——1867, Geology of Antigonish County, Nova Scotia: Nova Scotia Inst. Nat. Sci., v. I, pt. IV, p. 106–120.

——1870, Nova Scotian geology: Nova Scotia Inst. Nat. Sci., v. II, pt. 4, p. 67.

——1886, A revision of the geology of Antigonish County, Nova Scotia: Nova Scotia Inst. Nat. Sci., v. VI, p. 308–325.

——1890, Notes of examination by Professor James Hall of the Silurian collection of the Provincial Museum: Nova Scotia Inst. Nat. Sci. Trans., v. VII, p. 14–17.

Hubbert, M. K., 1951, Mechanical basis for certain familiar geological structures: Geol. Soc. America Bull., v. 62, p. 355–372.

Jones, O. T., 1926, Notes on the Silurian rocks at Arisaig, Nova Scotia: Am. Jour. Sci., 5th ser., v. 11, p. 119–125.

Kaminsky, P. C., 1953, The geology of northern Cape George, Nova Scotia [M.S. thesis]: Cambridge, Massachusetts Inst. Technology.

Kelley, D. G., 1966, Cobequid Mountains, *in* Report of activities, May to October 1965, Jenness, S. E., ed., Canada Geol. Survey Paper 66–1, p. 172–173.

Kramer, J. R., and Phinney, W., 1953, The geology of Cape George, Nova Scotia [S.B. thesis]: Cambridge, Massachusetts Inst. Technology.

Kuenen, Ph. H., 1959, Fluviatile action on sand: Am. Jour. Sci., v. 257, p. 172–190.

——1964, Surf action: Sedimentology, v. 1, p. 29.

Legault, J., 1966, The Conodont fauna of the Stonehouse Formation, Arisaig, Nova Scotia [M.S. thesis]: Ottawa, Vanier Library, Ottawa Univ., 71 p., 6 pls., 3 figs.

Lindeman, E., and Bolton, L. L., 1917, Iron ore occurrences in Canada (2 Vols.): Canada Dept. Mines, Mines Branch, p. 222 plus 71 pl.

Maehl, R. H., 1961, The older Paleozoic of Pictou County, Nova Scotia: Province of Nova Scotia Dept. of Mines Mem. 4, 110 p.

McKee, E. D., 1938, Original structure in Colorado River flood deposits of Grand Canyon: Jour. Sed. Petrology, v. 8, p. 77–83.

——1939, Some types of bedding in the Colorado River delta: Jour. Geology, v. 47, p. 64–81.

McKerrow, W. S., and Ziegler, A. M., 1971, The Lower Silurian paleogeography of New Brunswick and adjacent areas: Jour. Geology, v. 79, p. 635–646.

McLearn, F. H., 1918, The Silurian Arisaig Series of Arisaig, Nova Scotia: Am. Jour. Sci., v. 45, p. 126–140.

——1924, Paleontology of the Silurian rocks of Arisaig, Nova Scotia: Canada Geol. Survey Mem. 137, 180 p.

Murray, Bruce, 1956, Stratigraphy of the Horton Group in parts of Nova Scotia [Ph.D. thesis]: Cambridge, Massachusetts Inst. Technology.

Naylor, R. S., and Boucot, A. J., 1965, Origin and distribution of rocks of Ludlow age (late Silurian) in the northern Appalachians: Am. Jour. Sci., v. 263, p. 153–169.

Ramsay, J. G., 1962, The geometry of conjugate fold systems; Geol. Mag., v. 93, p. 345–350.

Rast, N., 1963, Textural evidence for the origin of ignimbrites: Geol. Jour., v. 3, p. 97–108.

Rexroad, C. B., Branson, E. R., Smith, M. O., Summerson, C., and Boucot, A. J., 1965, The Silurian formations of east-central Kentucky and adjacent Ohio: Kentucky Geol. Survey Bull., v. 2, p. 1–34.

Sage, N. Mcl., Jr., 1954, The stratigraphy of the Windsor Group in the Antigonish quadrangles and the Mahone Bay, St. Margaret Bay area: Nova Scotia Dept. Mines Mem. 3, 168 p.

Sheldon, R. P., 1964, Barrier islands and lagoonal iron sedimentation in the Silurian of Alabama [abs.]: Geol. Soc. America Spec. Paper 82, p. 182.

Smitheringale, W. G., 1960, Geology of Nictaux-Torbrook map-area, Annapolis and Kings Counties, Nova Scotia: Canada Geol. Survey Paper 60–13, p. 1–32.

Stacy, Maurice E., 1953, Stratigraphy and paleontology of the Windsor Group (Upper Mississippian) in parts of Cape Breton Island, Nova Scotia: Province of Nova Scotia Dept. of Mines, Mem. 2.

Sykes, G., 1937a, The Colorado Delta: Carnegie Inst. Washington Pub. 460.

——1937b, Delta, estuary and lower portion of the channel of the Colorado River, 1933 to 1935: Carnegie Inst. Washington Pub. 480.

Twenhofel, W. H., 1909, The Silurian section at Arisaig, Nova Scotia: Am. Jour. Sci., v. 4, no. 28, p. 143–169.

Wait, J. H., 1959, Geology of the Mabou area, Cape Breton Island, Nova Scotia, Canada [B.S. thesis]: Lewiston, Maine, Bates College.

Walliser, O. H., 1957, Conodonten aus dem oberen Gotlandium Deutschlands und der Karnischen Alpen: Notizbl. Hess. Landesamt Bodenf., Bd. 85, p. 28–52.

Weeks, L. J., 1954, Southeast Cape Breton Island, Nova Scotia: Canada Geol. Survey Mem. 277, p. 1–112.

White, E. I., 1935, The ostracoderm *Pteraspis* Kier and the relationships of the agnathous vertebrates: Royal Soc. London Philos. Trans., v. (B) 25, p. 381–457.

——1957, XI. Original environment of craniates, *in* Westoll, T. S., ed., Studies on fossil vertebrates: London, p. 212–234.

Williams, A., 1951, Llandovery brachiopods from Wales with special reference to the Llandovery district: Geol. Soc. London Quart. Jour., v. 107, p. 85–136.

Williams, M. Y., 1914, Arisaig-Antigonish district, Nova Scotia: Canada Geol. Survey Mem. 60, 173 p.

Wilson, J. T., 1963, Continental drift: Sci. American, v. 208 (4), p. 86–100.

Ziegler, A. M., 1965, Silurian marine communities and their environmental significance: Nature, v. 207, no. 4994, p. 270–272.

——1966, The Silurian brachiopod *Eocoelia hemisphaerica* (J. de C. Sowerby) and related species: Palaeontology, v. 9, p. 523–543, Pls. 83–84.

Ziegler, A. M., Cocks, L.R.M., and Bambach, R. K., 1968, The composition and structure of Lower Silurian marine communities: Lethaia, v. 1, no. 1, p. 1–27.

MANUSCRIPT RECEIVED BY THE SOCIETY APRIL 25, 1967
REVISED MANUSCRIPT RECEIVED FEBRUARY 22, 1971

Index

Index